Finite Element Methods for Viscous Incompressible Flows

A Guide to Theory, Practice, and Algorithms

This is a volume in
COMPUTER SCIENCE AND SCIENTIFIC COMPUTING

Werner Rheinboldt and Daniel Siewiorek, editors

A list of titles in this series appears at the end of this volume.

Finite Element Methods for Viscous Incompressible Flows

A Guide to Theory, Practice, and Algorithms

Max D. Gunzburger

Department of Mathematics
Carnegie Mellon University
Pittsburgh, Pennsylvania

Department of Mathematics
Virginia Polytechnic Institute and
State University
Blacksburg, Virginia

ACADEMIC PRESS, INC.
Harcourt Brace Jovanovich, Publishers

Boston San Diego New York
Berkeley London Sydney
Tokyo Toronto

ACADEMIC PRESS, INC.
1250 Sixth Avenue, San Diego, CA 92101

United Kingdom Edition published by
ACADEMIC PRESS INC. (LONDON) LTD.
24–28 Oval Road, London NW1 7DX

Library of Congress Cataloging-in-Publication Data

Gunzburger, Max D.
  Finite element methods for viscous incompressible flows : a guide
to theory, practice, and algorithms / Max D. Gunzburger.

    p.   cm.—(Computer science and scientific computing)
    Bibliography: p.
    Includes indexes.
    ISBN 0-12-307350-2
    1. Viscous flow. 2. Finite element method. I. Title.
II. Series.
TA357.G86 1989
532'.052—dc19   89-353
                                                              CIP

Printed in the United States of America
89 90 91 92   9 8 7 6 5 4 3 2 1

To my parents

Table of Contents

Part II
SOLUTION OF THE DISCRETE EQUATIONS

Part V
THE STREAMFUNCTION FORMULATION

Part VI
EIGENVALUE PROBLEMS CONNECTED WITH STABILITY STUDIES FOR VISCOUS FLOWS

Part VII
EXTERIOR PROBLEMS

Part VIII
NONLINEAR CONSTITUTIVE RELATIONS

Part IX
ELECTROMAGNETICALLY OR THERMALLY COUPLED FLOWS

Part X
REMARKS ON SOME TOPICS THAT HAVE NOT BEEN CONSIDERED

Preface

A successful computational simulation of a natural process requires the selection of a mathematical model describing the process, a discretization algorithm that reduces the problem to one a computer can handle, and a method via which a solution of the discretized equations is found. The natural process considered here is the incompressible flow of a viscous fluid, or, as my daughter Cecilia puts it—and at times I agree with her assessement—incomprehensible vicious flow. The discretization techniques studied are finite element methods. One of the most successful and well developed mathematical theories concerning finite element methods is that connected with incompressible viscous flow problems. The success of this theory lies not only in the accumulation of elegant mathematical results, but also in its impact on practical computations.

A variety of mathematical models are treated in this book; the particular model one chooses to use depends very much on the specific application one has in mind. The central subject of the book is finite element discretization algorithms for mathematical models of various aspects of incompressible viscous flows. Results are given concerning the stability of the algorithms and the accuracy of approximate solutions obtained through their use. In addition, questions related to the implementation of these algorithms, including methods for the solution of the discrete equations, are also addressed.

In writing this book, I had three types of readers in mind. First, there are those who know little about the subject matter and want to get a quick taste of the main points and of the variety of results available. Then there are those who are experts in some phase of finite element methods for incompressible flows who want to learn something about other phases. Finally, there are experts in related fields such as the analysis of the Navier–Stokes equations or other numerical methods for incompressible flow problems who want to see what finite element methods have to offer. I hope these readers like what they see here and decide to explore the subject in greater depth by consulting the cited references.

It is presumed the reader has at least a rudimentary familiarity with the finite element method for the approximation of solutions of second-order elliptic partial differential equations. This material was not included because I wanted to keep the book short and accessible, not long and formidable. Furthermore, due to the availability of many good books covering this background material, the job has been more than adequately done for me. For example, one may consult any of Axelsson and Barker [1984], Babuska and Aziz [1972], Ciarlet [1978], Cuvelier, Segal, and van Steenhoven [1986], Johnson [1987], Oden and Reddy [1976], Strang and Fix [1973], Wait and Mitchell [1985], and Zienkiewicz [1977].

For those who want to delve more deeply into the subject, there exist various books, or parts of books, devoted to finite element methods for incompressible viscous flow problems. The outstanding monographs of Girault and Raviart [1979 and 1986] give a rigorous and detailed account of the mathematical theory and to this day remain the definitive sources. These books also contain lots of material that is of use to practitioners. Other books deal with important issues concerning the implementation of finite element algorithms; see, e.g., Cuvelier, Segal, and van Steenhoven [1986] and Thomasset [1981]. Also, various aspects of the finite element methodology are considered within Baker [1983], Chung [1978], Glowinski [1984], Glowinski, Lions, and Tremolieres [1981], and Peyret and Taylor [1983]. However, I should point out that the subjects considered in this book continue to evolve and much of the detailed source material can be found only in journals.

A principal goal is to present some of the important mathematical results that are relevant to practical computations. In so doing, useful algorithms are also discussed. Although rigorous results are stated, no detailed proofs are supplied; rather, the intention is to present these results so that they can serve as a guide for the selection and, in certain respects, the implementation of algorithms. Also, although numerous algorithms are examined, no specific numerical examples are presented. In short, I hope this book will be useful to mathematicians who want to know about some of the rigorous results known about the subject; if sufficiently interested, they can then look in the cited references for details. At the same time, I hope that the discussions of algorithms and their implementations will be useful to engineers and others involved in practical computations. These researchers should also be able to derive applicable information from some of the rigorous mathematical results discussed.

The core of the book is Parts I–V. Different mathematical models, e.g., primitive variable, streamfunction–vorticity, etc., of incompressible viscous flows are considered. The implementation of a variety of different boundary conditions is a topic of continuing interest. Both the discretization and solution processes are discussed.

Parts VI–IX are concerned with variations on the themes of the first five parts, i.e., with a variety of what may be considered special topics involving incompressible viscous flows. My purpose is to examine the wide applicability of finite element methods to these different situations.

Although I have attempted to be broad and broad minded in my selection of topics, space limitations have forced me to leave out some important items. Some of these are briefly mentioned in the last part; some are surely left out altogether.

Acknowledgments

My own work on the subject of this book, as well as that of my students, has benefited from the financial support of various organizations. Foremost among these is the Air Force Office of Scientific Research under grants AFOSR-80-0083, AFOSR-83-0101, and AFOSR-88-0197. I want to express my sincere appreciation to that organization and to the monitors of these grants. Thanks are also due to the Institute for Computer Applications in Science and Engineering, to the Los Alamos National Laboratory, to Virginia Polytechnic Institute and State University, to the University of Tennessee, and especially, to Carnegie Mellon University.

My work has benefited even more from discussions and collaborations with many friends, students, teachers, and colleagues. I especially wish to acknowledge the many fruitful interactions, both of substance and of spirit, with George Fix, Nic Nicolaides, and Janet Peterson.

I

Discretizations of the Primitive Variable Formulation

We begin our excursion with the simplest model of incompressible viscous flow, which retains the two features that are the most troublesome to the design, analysis and implementation of finite element algorithms. The model is the primitive variable, i.e., velocity–pressure, formulation with homogeneous velocity boundary conditions, and the troublesome features are the incompressibility constraint and the nonlinearity of the convection term. Following a detailed discussion of finite element approximations of this model, we consider other primitive variable formulations of incompressible viscous flow problems, especially as they concern other boundary conditions.

1

A Primitive Variable Formulation

Let Ω denote an open, bounded, possibly multiply connected, domain in $\mathbb{R}^n$, $n = 2$ or 3, and let Γ denote its boundary. As a prototype for incompressible flow problems, we consider the Navier–Stokes equations,

$$-\nu \Delta \mathbf{u} + \mathbf{u} \cdot \operatorname{grad} \mathbf{u} + \operatorname{grad} p = \mathbf{f} \quad \text{in } \Omega, \tag{1.1}$$

together with the incompressibility constraint,

$$\operatorname{div} \mathbf{u} = 0 \quad \text{in } \Omega, \tag{1.2}$$

and the homogeneous no-slip boundary condition,

$$\mathbf{u} = 0 \quad \text{on } \Gamma, \tag{1.3}$$

where $\mathbf{u}$ denotes the velocity field, p the pressure, $\mathbf{f}$ the given body force per unit mass, and ν the given constant kinematic viscosity. In (1.1) the constant density has been absorbed into the pressure. Whenever $\mathbf{u}$ and p represent nondimensional variables, ν is the inverse of the Reynolds number Re. There are a vast number of sources that may be consulted for the derivation of (1.1)–(1.3) from first principles and for classical results about properties of their solutions. See, for example, Landau and Lifshitz [1987] and Serrin [1959].

1.1. Function Spaces, Norms, and Forms

In order to introduce a Galerkin-type weak formulation through which
a finite element approximation is determined, we first need to define
some function spaces and associated norms, and also forms involving
functions belonging to those spaces. More detailed accounts concerning
these spaces may be found in, e.g., Adams [1975], Duvaut and Lions
[1976], Grisvard [1985], Necas [1967] and Wloka [1987]; Baiocchi and
Capelo [1984] and Girault and Raviart [1986] provide lucid and concise
expositions.

First we denote by $L^2(\Omega)$ the space of functions that are square integrable over Ω and that are equipped with the inner product and norm

$$(p, q) = \int_\Omega pq \, d\Omega \quad \text{and} \quad \|q\|_0 = (q, q)^{1/2},$$

respectively. We then define the constrained space

$$L_0^2(\Omega) = \left\{ q \in L^2(\Omega) : \int_\Omega q \, d\Omega = 0 \right\}.$$

Thus $L_0^2(\Omega)$ consists of square integrable functions having zero mean
over Ω. This space is used in connection with the pressure; the constraint is needed since it is clear from (1.1)–(1.3) that the pressure can
be determined only up to an arbitrary additive constant. In practice,
other constraints, e.g., fixing the pressure at a point, may be used
instead without effecting any appreciable change in the algorithms
discussed below.

Next, for any non-negative integer k, we define the Sobolev space

$$H^k(\Omega) = \{q \in L^2(\Omega) : D^s q \in L^2(\Omega) \quad \text{for } s = 1, ..., k\},$$

where D^s denotes any and all derivatives of order s. Thus $H^k(\Omega)$ consists of square integrable functions all of whose derivatives of order up
to k are also square integrable. $H^k(\Omega)$ comes equipped with the norm

$$\|q\|_k = (\|q\|_0^2 + \sum \|D^s q\|_0^2)^{1/2},$$

where the summation extends over all possible derivatives of order k or
less. Clearly $H^0(\Omega) = L^2(\Omega)$. Of particular interest is the space $H^1(\Omega)$
consisting of functions with one square integrable derivative and the
subspace

$$H_0^1(\Omega) = \{q \in H^1(\Omega) : q = 0 \text{ on } \Gamma\}$$

whose elements have one square integrable derivative over Ω and vanish on the boundary Γ. These spaces have the associated norm

$$\|q\|_1 = \left(\|q\|_0^2 + \sum_{i=1}^n \left\| \frac{\partial q}{\partial x_i} \right\|_0^2 \right)^{1/2}. \tag{1.4}$$

For functions belonging to $H_0^1(\Omega)$ the semi-norm

$$|q|_1 = \left(\sum_{i=1}^n \left\| \frac{\partial q}{\partial x_i} \right\|_0^2 \right)^{1/2} \tag{1.5}$$

defines a norm equivalent to (1.4) and thus, for such functions, (1.5) may be used instead of (1.4).

For noninteger $k > 0$, we may define the space $H^k(\Omega)$ through a well-known interpolation process. Also, we denote by $H^{-1}(\Omega)$ the dual space consisting of bounded linear functionals on $H_0^1(\Omega)$, i.e., $q \in H^{-1}(\Omega)$ implies that $(q, v) < \infty$ for all $v \in H_0^1(\Omega)$. A norm for $H^{-1}(\Omega)$ is given by

$$\|q\|_{-1} = \sup_{0 \neq v \in H_0^1(\Omega)} \frac{(q, v)}{|v|_1}.$$

We will also use the trace spaces, which consist of the restriction, to the boundary Γ, of functions belonging to $H^k(\Omega)$. For example, $H^{1/2}(\Gamma)$ consists of traces of functions belonging to $H^1(\Omega)$; a norm for functions belonging to $H^{1/2}(\Gamma)$ may be defined by

$$\|q\|_{1/2,\Gamma} = \inf_{\substack{v \in H^1(\Omega) \\ v = q \text{ on } \Gamma}} \|v\|_1.$$

For vector valued functions we use the spaces

$$\mathbf{H}^k(\Omega) = [H^k(\Omega)]^n = \{\mathbf{v} : v_i \in H^k(\Omega) \quad \text{for } i = 1, ..., n\},$$

$$\mathbf{H}_0^1(\Omega) = [H_0^1(\Omega)]^n = \{\mathbf{v} : v_i \in H_0^1(\Omega) \quad \text{for } i = 1, ..., n\},$$

$$\mathbf{H}^{-1}(\Omega) = [H^{-1}(\Omega)]^n = \{\mathbf{v} : v_i \in H^{-1}(\Omega) \quad \text{for } i = 1, ..., n\},$$

and

$$\mathbf{H}^{1/2}(\Gamma) = [H^{1/2}(\Gamma)]^n = \{\mathbf{v} : v_i \in H^{1/2}(\Gamma) \quad \text{for } i = 1, ..., n\}.$$

For example, $\mathbf{H}^k(\Omega)$ consists of vector valued functions each of whose components belongs to $H^k(\Omega)$. For integer $k \geq 0$, $\mathbf{H}^k(\Omega)$ is equipped with the norm

$$\|\mathbf{v}\|_k = \left(\sum_{i=1}^n \|v_i\|_k^2 \right)^{1/2};$$

alternately, for functions belonging to $\mathbf{H}_0^1(\Omega)$, one may use

$$|\mathbf{v}|_1 = \left(\sum_{i=1}^n |v_i|_1^2 \right)^{1/2}.$$

Extensions of the definitions of the norms associated with $\mathbf{H}^k(\Omega)$ for noninteger and negative k are also possible. Also, the inner product for functions belonging to $\mathbf{L}^2(\Omega) = \mathbf{H}^0(\Omega) = [L^2(\Omega)]^n$ is given by

$$(\mathbf{u}, \mathbf{v}) = \int_\Omega \mathbf{u} \cdot \mathbf{v} \, d\Omega,$$

where there is no ambiguity possible resulting from using the same notation for both the inner products of scalar and vector valued functions.

We now define the bilinear forms

$$a(\mathbf{u}, \mathbf{v}) = \nu \int_\Omega \operatorname{grad} \mathbf{u} : \operatorname{grad} \mathbf{v} \, d\Omega \quad \text{for all } \mathbf{u}, \mathbf{v} \in \mathbf{H}^1(\Omega), \tag{1.6}$$

$$b(\mathbf{v}, q) = -\int_\Omega q \operatorname{div} \mathbf{v} \, d\Omega \quad \text{for all } \mathbf{v} \in \mathbf{H}^1(\Omega) \quad \text{and} \quad q \in L^2(\Omega), \tag{1.7}$$

and the trilinear form

$$c(\mathbf{w}, \mathbf{u}, \mathbf{v}) = \int_\Omega \mathbf{w} \cdot \operatorname{grad} \mathbf{u} \cdot \mathbf{v} \, d\Omega \quad \text{for all } \mathbf{u}, \mathbf{v}, \mathbf{w} \in \mathbf{H}^1(\Omega). \tag{1.8}$$

In (1.6) and (1.8) we have that $(\operatorname{grad} \mathbf{u})_{ij} = \partial u_j / \partial x_i$,

$$\operatorname{grad} \mathbf{u} : \operatorname{grad} \mathbf{v} = \sum_{i,j=1}^n \frac{\partial u_i}{\partial x_j} \frac{\partial v_i}{\partial x_j},$$

and

$$\mathbf{w} \cdot \operatorname{grad} \mathbf{u} \cdot \mathbf{v} = \sum_{i,j=1}^n w_j \frac{\partial u_i}{\partial x_j} v_i.$$

Using the bilinear form $b(\cdot, \cdot)$, we can define the subspace

$$\mathbf{Z} = \{\mathbf{v} \in \mathbf{H}_0^1(\Omega) : b(\mathbf{v}, q) = 0 \quad \text{for all } q \in L_0^2(\Omega)\},$$

which consists of (weakly) *divergence free functions*, i.e., functions whose divergence vanishes almost everywhere. Certainly any function that is divergence free in the strong sense belongs to $\mathbf{Z}$.

1.2. A Galerkin-Type Weak Formulation

The most commonly used weak formulation of (1.1)–(1.3) is the following. Given $\mathbf{f} \in \mathbf{H}^{-1}(\Omega)$, we seek $\mathbf{u} \in \mathbf{H}_0^1(\Omega)$ and $p \in L_0^2(\Omega)$ such that

$$a(\mathbf{u}, \mathbf{v}) + c(\mathbf{u}, \mathbf{u}, \mathbf{v}) + b(\mathbf{v}, p) = (\mathbf{f}, \mathbf{v}) \qquad \text{for all } \mathbf{v} \in \mathbf{H}_0^1(\Omega) \qquad (1.9)$$

and

$$b(\mathbf{u}, q) = 0 \qquad \text{for all } q \in L_0^2(\Omega). \qquad (1.10)$$

By virtue of (1.10) we see that the velocity $\mathbf{u}$ belongs to $\mathbf{Z}$.

It can be easily verified that whenever a pair $(\mathbf{u}, p)$ satisfies (1.9)–(1.10) and is sufficiently smooth to allow for the appropriate integrations by parts, then $(\mathbf{u}, p)$ is also a solution of (1.1)–(1.3). Of course, (1.9)–(1.10) admit solutions that are not sufficiently smooth to be solutions of (1.1)–(1.3); hence the terminologies *weak formulation* and *generalized solutions* are applied to (1.9)–(1.10) and their solutions, respectively. On the other hand, it is also clear that any solution of (1.1)–(1.3), i.e., any *strong solution*, satisfies (1.9)–(1.10).

For the weak formulation (1.9)–(1.10), the boundary condition (1.3) is an *essential* one, i.e., it must be imposed on the candidate solution functions. Below, in Section 4.3, we will discuss the *natural* boundary conditions associated with the weak formulation (1.9)–(1.10).

We will not enter into details concerning the existence, uniqueness, continuous dependence on data, and regularity of solutions of (1.9)–(1.10). Such results may be found in, e.g., Girault and Raviart [1979 and 1986], Ladyzhenskaya [1969], and Temam [1979 and 1983]. Futhermore, many of these results are similar to those discussed below for the approximate problem.

2

The Finite Element Problem and the Div-Stability Condition

2.1. The Discrete Finite Element Problem

Once the Galerkin formulation (1.9)–(1.10) is established, the approximate problem from which the finite element solution is determined is defined in the usual manner. First one chooses the approximating finite element spaces, or more precisely, a family of finite element spaces, $\mathbf{V}_0^h$ and S_0^h for the velocity and pressure, respectively. Here h is a parameter that is usually related to the size of the grid associated with the subdivision of $\bar{\Omega}$ into finite elements. Then one requires that (1.9)–(1.10) hold for functions belonging to these finite dimensional spaces, i.e., one seeks $\mathbf{u}^h \in \mathbf{V}_0^h$ and $p^h \in S_0^h$ such that

$$a(\mathbf{u}^h, \mathbf{v}^h) + c(\mathbf{u}^h, \mathbf{u}^h, \mathbf{v}^h) + b(\mathbf{v}^h, p^h) = (\mathbf{f}, \mathbf{v}^h) \qquad \text{for all } \mathbf{v}^h \in \mathbf{V}_0^h \quad (2.1)$$

and

$$b(\mathbf{u}^h, q^h) = 0 \qquad \text{for all } q^h \in S_0^h. \tag{2.2}$$

If $\mathbf{V}_0^h$ and S_0^h are subspaces of the underlying infinite dimensional spaces used in (1.9)–(1.10), i.e., if $\mathbf{V}_0^h \subset \mathbf{H}_0^1(\Omega)$ and $S_0^h \subset L_0^2(\Omega)$, then the finite element method defined by (2.1)–(2.2) is said to be *conforming*. Otherwise, i.e., if $\mathbf{V}_0^h \not\subset \mathbf{H}_0^1(\Omega)$ and/or $S_0^h \not\subset L_0^2(\Omega)$, then the method is said to be *nonconforming*. We will restrict our attention to examples of the former.

Once one chooses specific bases for $\mathbf{V}_0^h$ and S_0^h, (2.1)–(2.2) are equivalent to a *nonlinear system of algebraic equations*. Indeed, if $\{q_j(\mathbf{x})\}$, $j = 1, ..., J$ and $\{\mathbf{v}_k(\mathbf{x})\}$, $k = 1, ..., K$, denote basis sets for S_0^h and $\mathbf{V}_0^h$, respectively, we may then write

$$p^h = \sum_{j=1}^{J} \alpha_j q_j(\mathbf{x}) \quad \text{and} \quad \mathbf{u}^h = \sum_{k=1}^{K} \beta_k \mathbf{v}_k(\mathbf{x})$$

for some constants $\alpha_j, j = 1, ..., J$, and $\beta_k, k = 1, ..., K$. The substitution of these expressions into (2.1)–(2.2) yields

$$\sum_{k=1}^{K} a(\mathbf{v}_k, \mathbf{v}_\ell)\beta_k + \sum_{k,m=1}^{K} c(\mathbf{v}_m, \mathbf{v}_k, \mathbf{v}_\ell)\beta_k \beta_m + \sum_{j=1}^{J} b(\mathbf{v}_\ell, q_j)\alpha_j = (\mathbf{f}, \mathbf{v}_\ell)$$

$$\text{for } \ell = 1, ..., K \tag{2.3}$$

and

$$\sum_{k=1}^{K} b(\mathbf{v}_k, q_i)\beta_k = 0 \quad \text{for } i = 1, ..., J, \tag{2.4}$$

which constitute a nonlinear algebraic, in fact, quadratic, system of $J + K$ equations for the $J + K$ unknowns $\alpha_j, j = 1, ..., J$, and β_k, $k = 1, ..., K$. Note that the discrete continuity equation (2.2) yields the $J \times K$ rectangular *linear* system (2.4).

2.2. The Div-Stability Condition

In the positive definite case, e.g., for the equations of linear elasticity, the mere inclusion of the finite element spaces within the underlying function spaces is essentially sufficient to assure that the approximate solution is well defined and is, as far as the rate of convergence is concerned, as accurate as possible for the type of finite element functions being used. Here, for the Navier–Stokes equations, the inclusions $\mathbf{V}_0^h \subset \mathbf{H}_0^1(\Omega)$ and $S_0^h \subset L_0^2(\Omega)$ are not by themselves sufficient to produce stable, meaningful approximations. We find ourselves in the realm of what are known as *mixed finite element methods*. Thus, there are a number of conditions that the finite element spaces should satisfy and, although most of these are satisfied by arbitrary choices of conforming finite element spaces, one of the conditions restricts this choice.

The four "easy" conditions that the finite element spaces $\mathbf{V}_0^h$ and S_0^h

and the forms (1.6)–(1.8) satisfy by virtue of the inclusions $\mathbf{V}_0^h \subset \mathbf{H}_0^1(\Omega)$ and $S_0^h \subset L_0^2(\Omega)$ are given as follows. First, we have the three continuity, or boundedness conditions: There exist positive constants κ_a, κ_b, and κ_c, independent of h, such that

$$|a(\mathbf{u}^h, \mathbf{v}^h)| \le \kappa_a |\mathbf{u}^h|_1 |\mathbf{v}^h|_1 \quad \text{for all } \mathbf{u}^h, \mathbf{v}^h \in \mathbf{V}_0^h, \qquad (2.5)$$

$$|b(\mathbf{v}^h, q^h)| \le \kappa_b |\mathbf{v}^h|_1 \|q^h\|_0 \quad \text{for all } \mathbf{v}^h \in \mathbf{V}_0^h \quad \text{and} \quad q^h \in S_0^h, \quad (2.6)$$

and

$$|c(\mathbf{w}^h, \mathbf{u}^h, \mathbf{v}^h)| \le \kappa_c |\mathbf{u}^h|_1 |\mathbf{v}^h|_1 |\mathbf{w}^h|_1 \quad \text{for all } \mathbf{u}^h, \mathbf{v}^h, \mathbf{w}^h \in \mathbf{V}_0^h. \quad (2.7)$$

These are valid since one may easily show that they hold over the entire spaces $\mathbf{H}_0^1(\Omega)$ and $L_0^2(\Omega)$.

The fourth condition makes use of the subspace

$$\mathbf{Z}^h = \{\mathbf{v}^h \in \mathbf{V}_0^h : b(\mathbf{v}^h, q^h) = 0 \quad \text{for all } q^h \in S_0^h\}$$

of *discretely divergence free functions* associated with the finite element spaces $\mathbf{V}_0^h$ and S_0^h and the bilinear form $b(\cdot, \cdot)$. In general, $\mathbf{Z}^h \not\subset \mathbf{Z}$, even when $\mathbf{V}_0^h \subset \mathbf{H}_0^1(\Omega)$ and $S_0^h \subset L_0^2(\Omega)$, i.e., discretely solenoidal functions are not necessarily solenoidal. This is, of course, entirely analogous to the finite difference case, e.g., a function satisfying a difference approximation to the incompressibility constraint is not, in general, solenoidal. A measure of the "angle" between the spaces $\mathbf{Z}^h$ and $\mathbf{Z}$ is given by

$$\Theta = \sup_{\substack{\mathbf{z}^h \in \mathbf{Z}^h \\ |\mathbf{z}^h|_1 = 1}} \inf_{\mathbf{z} \in \mathbf{Z}} |\mathbf{z} - \mathbf{z}^h|_1. \qquad (2.8)$$

In general, $0 \le \Theta \le 1$, which is easily seen by observing that $\Theta = 0$ whenever $\mathbf{Z}^h \subset \mathbf{Z}$ and $\Theta = 1$ whenever $\mathbf{z} = 0$. Thus, if it so happens that discretely divergence free functions are actually solenoidal, i.e., $\mathbf{Z}^h \subset \mathbf{Z}$, then $\Theta = 0$. It follows from (2.2) that the approximate velocity $\mathbf{u}^h \in \mathbf{Z}^h$, i.e., $\mathbf{u}^h$ is discretely solenoidal. However, since in general $\mathbf{Z}^h \not\subset \mathbf{Z}$, div $\mathbf{u}^h \neq 0$, even in a weak sense.

Having introduced the subspace $\mathbf{Z}^h$, the fourth "easy" condition is the (weak) coercivity condition: There exists a positive constant γ_a, independent of h, such that

$$\sup_{0 \neq \mathbf{v}^h \in \mathbf{Z}^h} \left(\frac{a(\mathbf{z}^h, \mathbf{v}^h)}{|\mathbf{v}^h|_1} \right) \ge \gamma_a |\mathbf{z}^h|_1 \quad \text{for all } \mathbf{z}^h \in \mathbf{Z}^h. \qquad (2.9)$$

This condition follows easily from the fact that $a(\mathbf{v}, \mathbf{v}) \ge \gamma_a |\mathbf{v}|_1^2$ for all $\mathbf{v} \in \mathbf{H}_0^1(\Omega)$.

The one condition that poses a problem when one chooses finite element spaces for Navier–Stokes computations has the following mathematical realization:

Given any $q^h \in S_0^h$,

$$\sup_{0 \neq \mathbf{v}^h \in \mathbf{V}_0^h} \left(\frac{b(\mathbf{v}^h, q^h)}{|\mathbf{v}^h|_1} \right) \geq \gamma \|q^h\|_0, \tag{2.10}$$

where the constant $\gamma > 0$ may be chosen independent of h and of the particular choice of $q^h \in S_0^h$.

This condition may be equivalently expressed in the form:

Given any $q^h \in S_0^h$ there exists a nonzero $\mathbf{v}^h \in \mathbf{V}_0^h$ such that

$$b(\mathbf{v}^h, q^h) \geq \gamma \|q^h\|_0 |\mathbf{v}^h|_1, \tag{2.11}$$

where the constant $\gamma > 0$ may be chosen independent of h and of the particular choice of $q^h \in S_0^h$.

Of course, for each q^h a different $\mathbf{v}^h$ may be chosen in order to satisfy (2.11).

The condition (2.10), or equivalently (2.11), is variously known as the *Ladyzhenskaya-Babuska-Brezzi* (or *LBB*) or the *inf-sup* condition, the latter designation following from the third equivalent form:

There exists a $\gamma > 0$, independent of h, such that

$$\inf_{0 \neq q^h \in S_0^h} \sup_{0 \neq \mathbf{v}^h \in \mathbf{V}_0^h} \left(\frac{b(\mathbf{v}^h, q^h)}{|\mathbf{v}^h|_1 \|q^h\|_0} \right) \geq \gamma. \tag{2.12}$$

The LBB designation is a result of the fact that (2.10) is an example of the stability condition found in the Brezzi theory for mixed finite element methods (Brezzi [1974]; see also Babuska [1973]), which in turn is essentially an application of the Babuska theory for finite element methods (Babuska [1971] and Babuska and Aziz [1972]), and also from the fact that Ladyzhenskaya proved the analogous condition for the continuous case, i.e., on $\mathbf{H}_0^1(\Omega) \times L_0^2(\Omega)$ (Ladyzhenskaya [1969]).

Following Boland and Nicolaides [1983], we will refer to any of the equivalent statements (2.10)–(2.12) as the condition for *div-stability*. Note that the div-stability condition has nothing to do with the nonlinearity of the Navier–Stokes equations and, in fact, the possible problems its satisfaction poses are shared by the linear equations of Stokes flow.

We again note that due to (2.2), the approximate velocity $\mathbf{u}^h \in \mathbf{Z}^h$, i.e., $\mathbf{u}^h$ is discretely solenoidal. However, since in general $\mathbf{Z}^h \not\subset \mathbf{Z}$, div $\mathbf{u}^h \neq 0$. Loosely speaking, the div-stability condition (2.10) ensures, as $h \to 0$ at least, that discretely solenoidal functions tend to solenoidal functions. Below we will also be interested in the special cases of finite element spaces $\mathbf{V}_0^h$ and S_0^h for which $\mathbf{Z}^h \subset \mathbf{Z}$ so that (2.2) then implies that div $\mathbf{u}^h = 0$ almost everywhere.

It is possible to define meaningful approximations even when the finite element spaces $\mathbf{V}_0^h$ and S_0^h do not satisfy the div-stability condition; see Section 3.5 and Chapter 5. However, except in those instances, we will keep our discussion within the framework provided by the div-stability condition.

Those more familiar with finite difference methods for the approximate solution of the Navier–Stokes equations may wonder whether or not there is a div-stability condition present in such discretizations. The answer is yes. It is well known that if one uses central difference quotients, based on the same grid, to approximate all derivatives appearing in (1.1) and (1.2), then the resulting finite difference scheme is unstable. Through the use of physical intuition, this has led to the development of finite difference methods that employ, for instance, *staggered meshes* for defining the pressure and velocity unknowns; see Roache [1972]. Staggered meshes yield schemes that satisfy, in some sense, the div-stability condition.

2.3. Error Estimates and Other Results Concerning the Approximate Solution

We now present some of the available mathematical results concerning the solution $\mathbf{u}^h$, p^h of the finite element problem (2.1)–(2.2). Here we *assume* that the chosen finite element spaces $\mathbf{V}_0^h$ and S_0^h satisfy the div-stability condition (2.10) as well as the more transparent conditions (2.5)–(2.7) and (2.9). Subsequently, we will look into the issue of verifying the condition (2.10). The summary presented here is based on the detailed analyses found in Brezzi, Rappaz, and Raviart [1980], Crouzeix and Raviart [1973], Girault and Raviart [1979 and 1986], Gunzburger and Peterson [1983], Jamet and Raviart [1973], and Temam [1979].

First, for any $\mathbf{f} \in \mathbf{H}^{-1}(\Omega)$, (2.1)–(2.2) have a solution $\mathbf{u}^h$, p^h, provided that the div-stability condition (2.10) holds. However, one can prove that the solution is unique only for "sufficiently small" data $\mathbf{f}$ or "sufficiently large" viscosity v. More precisely, it is well known that the condition

$$\frac{\kappa_c \|\mathbf{f}\|_{-1}}{v^2} < 1 \qquad (2.13)$$

guarantees the uniqueness of the solution of (1.9)–(1.10). (Of course, if (2.13) does not hold, one cannot infer any information concerning the uniqueness, or lack thereof, of that solution.) One can then show that whenever (2.13) and the div-stability condition (2.10) hold, then for sufficiently small h the discrete solution of (2.1)–(2.2) is also uniquely determined.

If one can show that (1.9)–(1.10) have a unique solution, it can also be shown that the finite element solution of (2.1)–(2.2) converges to that solution. In addition, something can be said about the convergence of the finite element solution even when (1.9)–(1.10) do not possess a unique solution.

Briefly, we introduce the notion of a *branch of nonsingular solutions of* (1.9)–(1.10); details concerning this concept can be found in Brezzi, Rappaz, and Raviart [1980] and Girault and Raviart [1986]. Recall that whenever $\mathbf{u}$ and p represent nondimensional variables, then the Reynolds number $Re = 1/v$. For a compact interval $\mathfrak{I} \subset \mathbb{R}^+$, we examine the branch of solutions $\{\mathbf{u}(Re), p(Re) : Re \in \mathfrak{I}\}$. This branch is called nonsingular if the *linear* problem

$$a(\mathbf{w}, \mathbf{v}) + c(\mathbf{u}, \mathbf{w}, \mathbf{v}) + c(\mathbf{w}, \mathbf{u}, \mathbf{v}) + b(\mathbf{v}, s) = (\mathbf{f}, \mathbf{v}) \qquad \text{for all } \mathbf{v} \in \mathbf{H}_0^1(\Omega)$$
$$(2.14)$$

and

$$b(\mathbf{w}, q) = 0 \qquad \text{for all } q \in L_0^2(\Omega) \qquad (2.15)$$

has a unique solution $\mathbf{w}, s \in \mathbf{H}_0^1(\Omega) \times L_0^2(\Omega)$ for every $\mathbf{f} \in \mathbf{H}^{-1}(\Omega)$ and any $\mathbf{u}$ on the branch. Note that (2.14)–(2.15) are merely a weak formulation of

$$-v \, \Delta \mathbf{w} + \mathbf{u} \cdot \text{grad } \mathbf{w} + \mathbf{w} \cdot \text{grad } \mathbf{u} + \text{grad } s = \mathbf{f} \text{ in } \Omega,$$

$$\text{div } \mathbf{w} = 0 \text{ in } \Omega \qquad \text{and} \qquad \mathbf{w} = 0 \text{ on } \Gamma,$$

namely, the Navier–Stokes equations linearized about $\mathbf{u}$. Essentially, saying that $(\mathbf{u}, p)$ belong to a nonsingular branch of solutions means that we are not at bifurcation points or turning points.

Provided the div-stability condition holds, it can be shown that (2.1)–(2.2) possess a branch of nonsingular solutions that, as $h \to 0$, converge to a given branch of nonsingular solutions of (1.9)–(1.10). Thus the convergence of finite element approximations can be guaranteed even for values of the Reynolds number for which the Navier–Stokes equations do not possess a unique solution, provided one stays away from turning and bifurcation points.

Error estimates can also be derived, either when uniqueness can be guaranteed or when the solution is known to belong to a nonsingular branch. In either case, provided that the div-stability condition is satisfied, we have that

$$|\mathbf{u} - \mathbf{u}^h|_1 \leq C_1 \inf_{\mathbf{v}^h \in \mathbf{V}_0^h} |\mathbf{u} - \mathbf{v}^h|_1 + C_2 \Theta \inf_{q^h \in S_0^h} \|p - q^h\|_0 \qquad (2.16)$$

and

$$\|p - p^h\|_0 \leq C_3 \inf_{\mathbf{v}^h \in \mathbf{V}_0^h} |\mathbf{u} - \mathbf{v}^h|_1 + C_4 \inf_{q^h \in S_0^h} \|p - q^h\|_0, \qquad (2.17)$$

where Θ is defined in (2.8) and C_i, $i = 1, ..., 4$, are constants independent of h. These estimates are optimal for the "graph norm" $|\mathbf{u}|_1 + \|p\|_0$ of functions belonging to $\mathbf{H}_0^1(\Omega) \times L_0^2(\Omega)$ in the sense that the rate of convergence of the finite element solution, measured in this norm, is the same as that of the best approximation to the pair $(\mathbf{u}, p)$ out of $\mathbf{V}_0^h \times S_0^h$.

It is of interest to examine the role that the div-stability condition (2.10) plays in the derivation of the error estimates (2.16) and (2.17). In fact, it is fairly easy to derive the estimate

$$|\mathbf{u} - \mathbf{u}^h|_1 \leq C_5 \inf_{\mathbf{z}^h \in \mathbf{Z}_0^h} |\mathbf{u} - \mathbf{z}^h|_1 + C_2 \Theta \inf_{q^h \in S_0^h} \|p - q^h\|_0 \qquad (2.18)$$

without invoking the div-stability condition. The first term on the right-hand side requires one to know how well one can approximate the divergence free function $\mathbf{u} \in \mathbf{Z}$ by elements belonging to $\mathbf{Z}^h$. It is here that the div-stability condition plays a crucial role; indeed, (2.10), along with the continuity condition (2.6), implies that

$$\inf_{\mathbf{z}^h \in \mathbf{Z}^h} |\mathbf{u} - \mathbf{z}^h|_1 \leq \left(1 + \frac{\kappa_b}{\gamma}\right) \inf_{\mathbf{v}^h \in \mathbf{V}_0^h} |\mathbf{u} - \mathbf{v}^h|_1 \qquad \text{for all } \mathbf{u} \in \mathbf{Z}. \quad (2.19)$$

Thus the div-stability condition ensures that divergence free functions can be approximated by discretely divergence free functions to essentially the same accuracy, as the former can be approximated by elements of $\mathbf{V}_0^h$. The combination of (2.18) and (2.19) then yields (2.17).

If the solution of (1.9)–(1.10), or more precisely, of the linearized adjoint problem corresponding to (1.9)–(1.10), is sufficiently regular, then one can obtain an improved velocity error estimate in the $\mathbf{L}^2(\Omega)$-norm, namely,

$$\|\mathbf{u} - \mathbf{u}^h\|_0 \le C_6 h |\mathbf{u} - \mathbf{u}^h|_1, \qquad (2.20)$$

where again C_6 is independent of h. For example, if $\Omega \subset \mathbb{R}^2$ is a bounded convex polygon, then (2.20) holds for div-stable piecewise linear finite element velocity spaces.

We see that once the div-stability condition is satisfied, the error in the finite element approximation depends only on the ability to approximate in the chosen finite element subspaces. In general, (2.16)–(2.17) indicate that the velocity and pressure errors are coupled. Furthermore, one finds that it is efficient to equilibrate the rates of convergence of the two terms on the right-hand sides of (2.16)–(2.17). For this reason, one would like to use, for example, polynomials of one degree higher for the velocity components than those used for the pressure. As a final comment, we note that the constants appearing in (2.16)–(2.17) are, in general, proportional to $1/\gamma$, where γ is the stability constant appearing in (2.10).

2.4. Verifying the Div-Stability Condition

For particular choices of $\mathbf{V}_0^h$ and S_0^h, it is usually *not* an easy matter to verify that the div-stability condition holds. To accomplish this task for families of such spaces is even more difficult. Here, we sketch four techniques for verifying the div-stability condition.

Fortin's Method. One seemingly attractive method of showing that the div-stability condition holds is due to Fortin [1977]. He has shown that the div-stability condition (2.10) is equivalent to the existence of a linear operator Π^h from $\mathbf{H}_0^1(\Omega) \to \mathbf{V}_0^h$ such that given any $\mathbf{v} \in \mathbf{H}_0^1(\Omega)$

$$b(\Pi^h \mathbf{v}, q^h) = b(\mathbf{v}, q^h) \qquad \text{for all } q^h \in S_0^h$$

and

$$|\Pi^h \mathbf{v}|_1 \le C |\mathbf{v}|_1,$$

where the constant $C > 0$ may be chosen independent of h and of the particular choice of $\mathbf{v} \in \mathbf{H}_0^1(\Omega)$. Thus the task of verifying the div-stability condition (2.10) is reduced to the task of showing the existence of the operator Π^h; unfortunately, although the latter task has been accomplished in a few specific settings, in general, it is also a difficult thing to do.

Verfurth's Method. Verfurth [1984a] has developed a method for verifying the div-stability condition (2.10) that applies to the case of *continuous* discrete pressure spaces. Specifically, if $S_0^h \subset H^1(\Omega) \cap L_0^2(\Omega)$, one starts out by combining the inverse inequality (see Ciarlet [1978])

$$|\mathbf{v}^h|_1 \leq C_1 h^{-1} \|\mathbf{v}^h\|_0 \quad \text{for all } \mathbf{v}^h \in \mathbf{V}_0^h, \tag{2.21}$$

with the result

$$\sup_{0 \neq \mathbf{v}^h \in \mathbf{V}_0^h} \frac{b(\mathbf{v}^h, q^h)}{\|\mathbf{v}^h\|_0} \geq C_2 |q^h|_1 \quad \text{for all } q^h \in S_0^h, \tag{2.22}$$

to yield

$$\sup_{0 \neq \mathbf{v}^h \in \mathbf{V}_0^h} \frac{b(\mathbf{v}^h, q^h)}{|\mathbf{v}^h|_1} \geq \frac{C_2}{C_1} h |q^h|_1 \quad \text{for all } q^h \in S_0^h. \tag{2.23}$$

The inequality (2.22) can be shown to hold for many element pairs involving continuous discrete pressure fields; see, e.g., Bercovier and Pironneau [1979]. Note that (2.22) has a similar appearance to the div-stability condition (2.10), but that it involves the "wrong" norms.

Next, one combines two additional inequalities. The first is that, given any $q^h \in S_0^h \subset L_0^2(\Omega)$, there exists a $\mathbf{w} \in \mathbf{H}_0^1(\Omega)$ such that $\operatorname{div} \mathbf{w} = q^h$ and

$$|\mathbf{w}|_1 \leq C_3 \|q^h\|_0; \tag{2.24}$$

see, e.g., Girault and Raviart [1979 and 1986], Ladyzhenskaya [1969], or Temam [1979]. The second results from invoking the following approximation theoretic assumption concerning the space $\mathbf{V}_0^h$: For any $\mathbf{w} \in \mathbf{H}_0^1(\Omega)$ there exists a $\mathbf{w}^h \in \mathbf{V}_0^h$ such that

$$\|\mathbf{w} - \mathbf{w}^h\|_0 + h|\mathbf{w} - \mathbf{w}^h|_1 \leq C_4 h |\mathbf{w}|_1. \tag{2.25}$$

Then, the combination of (2.24) and (2.25) yields

$$\sup_{0 \neq \mathbf{v}^h \in \mathbf{V}_0^h} \frac{b(\mathbf{v}^h, q^h)}{|\mathbf{v}^h|_1} \geq C_5 - C_6 h |q^h|_1 \quad \text{for all } q^h \in S_0^h \text{ with } \|q^h\|_0 = 1. \tag{2.26}$$

Verfurth then shows that the div-stability condition (2.10) follows from (2.23) and (2.26) provided the constants $C_1 \ldots, C_6$ are independent of h.

Thus the main task of applying his method, once the inverse inequality (2.21) and the approximation theoretic result (2.25) have been shown to hold for the discrete velocity space $\mathbf{V}_0^h$, is to show that (2.22) is valid.

The Boland–Nicolaides Method. A more useful method, in the sense of having wide applicability and relative ease of use, has been developed by Boland and Nicolaides [1983]. One difficulty with verifying the div-stability condition (2.10) is its *global* nature; Boland and Nicolaides have shown how to *localize* the difficult part of the verification process.

Specifically, consider a subdivision of Ω into disjoint *macro-elements* Ω_r, $r = 1, \ldots, R$, each of which consists of one or a few elements in the finite element triangulation associated with $\mathbf{V}_0^h$ and S_0^h. Let Γ_r denote the boundary of the macro-element Ω_r. See Figure 2.1 for an illustration. In general, not all macro-elements have to contain the same number of finite elements, but as the mesh size $h \to 0$, the number of different types of macro-elements must remain finite. Of greatest importance is that the number of finite elements within each type of macro-element is independent of h, i.e., as we refine the mesh, the macro-elements are also refined so that they always contain the same number of elements.

Boland and Nicolaides show that the div-stability condition (2.10) holds whenever the following two conditions are known to be true.

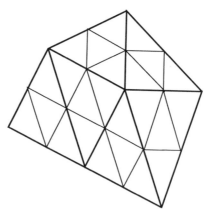

Figure 2.1

A patch of macro-elements, each consisting of four finite elements. _____ Macro-element boundaries, _____ finite element boundaries.

First, the div-stability condition should hold for the pair $\mathbf{V}_0^h$ and S_0^h *locally* over a macro-element, i.e., there exists a constant $\hat{\gamma} > 0$, independent of h and of the particular choice of macro-element, such that

$$\sup_{\mathbf{0} \neq \mathbf{v}^h \in \mathbf{V}_r^h} \frac{b(\mathbf{v}^h, q^h)}{|\mathbf{v}^h|_1} \geq \hat{\gamma} \|q^h\|_0 \quad \text{for all } q^h \in S_r^h, \tag{2.27}$$

where

$$\mathbf{V}_r^h = \{\mathbf{v} \in \mathbf{V}_0^h|_{\Omega_r} : \mathbf{v} = \mathbf{0} \text{ on } \Gamma_r\}, \quad \text{and} \quad S_r^h = \{q \in L_0^2(\Omega_r) \cap S_0^h|_{\Omega_r}\}, \tag{2.28}$$

where $L_0^2(\Omega_r)$ denotes the space of functions that are square integrable and have zero mean over Ω_r. Since $\mathbf{V}_r^h$ and S_r^h have fixed small dimension, independent of h, (2.27) may often be verified by a direct computation. The important observation is that (2.27) need be verified only once for each type of macro-element.

The second condition required in the Boland–Nicolaides method is that the div-stability condition should hold *globally* for the spaces $\tilde{\mathbf{V}}^h$ and $\tilde{S}^h$, where

$$\left.\begin{array}{l} \tilde{\mathbf{V}}^h \subset \mathbf{V}_0^h \\[2mm] \tilde{S}^h = \left\{\begin{array}{l} L_0^2(\Omega) \text{ piecewise constant functions with respect} \\ \text{to the macro-elements } \Omega_r, \, r = 1, \ldots, R \end{array}\right\} \end{array}\right\}, \tag{2.29}$$

i.e., there exists a constant $\tilde{\gamma} > 0$, independent of h, such that

$$\sup_{\mathbf{0} \neq \mathbf{v}^h \in \tilde{\mathbf{V}}^h} \frac{b(\mathbf{v}^h, q^h)}{|\mathbf{v}^h|_1} \geq \tilde{\gamma} \|q^h\|_0 \quad \text{for all } q^h \in \tilde{S}^h, \tag{2.30}$$

where $\tilde{\mathbf{V}}^h \subset \mathbf{V}_0^h$ and $\tilde{S}^h$ consists of functions that have zero mean over Ω and are constant in each macro-element Ω_r. If the requirements (2.27) and (2.30) are met, then the given spaces $\mathbf{V}_0^h$ and S_0^h are div-stable.

Summarizing the Boland–Nicolaides method, suppose we know that the pair $\mathbf{V}_0^h$, S_0^h is *locally div-stable* with constant $\hat{\gamma}$ independent of h, i.e., in the sense of (2.27). Further, suppose that the *comparison* spaces $\tilde{\mathbf{V}}^h$, $\tilde{S}^h$, which satisfy (2.29), are *globally div-stable* with constant $\tilde{\gamma}$ independent of h, i.e., in the sense of (2.30). Then the spaces $\mathbf{V}_0^h$, S_0^h are *globally div-stable* with a constant γ independent of h. Thus, through the use of the comparison spaces, the div-stability of the pair $\mathbf{V}_0^h$, S_0^h need only be checked locally, i.e., once for each type of macro-element.

This method has been successfully used, e.g., in Boland and Nicolaides [1983 and 1985] and Girault and Raviart [1986], to show the div-stability of a variety of known elements and some novel ones as well.

Stenberg's Method. Closely related to the method of Boland and Nicolaides is a method developed independently by Stenberg [1984]. We again introduce macro-elements and the local spaces (2.28), as in the Boland–Nicolaides method. Here, as is the case for the Boland–Nicolaides method, not all the macro-elements need be of the same type, e.g., there may be a different number of elements in different macro-elements. Of course, the number of different types of macro-elements should be finite as $h \to 0$. We then introduce the local null spaces

$$\mathbf{N}_r^h = \{q \in S_0^h|_{\Omega_r} : b(\mathbf{v}^h, q^h) = 0 \quad \text{for all } \mathbf{v}^h \in \mathbf{V}_r^h\}, \qquad (2.31)$$

$r = 1, \ldots, R$. Note that $\mathbf{N}_r^h$ is simply the null space of the discrete gradient operator restricted to the macro-element. Now suppose that for every macro-element Ω_r the space $\mathbf{N}_r^h$ is one dimensional, consisting of functions that are constant on Ω_r. If the velocity finite element space consists of piecewise linear or bilinear polynomials with respect to a triangulation of Ω into finite elements, we impose an additional geometric condition. This simply requires that the common part of the boundaries of two macro-elements is connected and contains at least two edges of the triangles or quadrilaterals of the finite element triangulation of $\bar{\Omega}$. (Figure 2.2 provides an illustration.) Stenberg shows that whenever these simple conditions hold, then the div-stability condition (2.10) holds as well.

Thus, Stenberg's method is very easy to apply. For quadratic, biquadratic, or higher order velocity spaces, one need only perform a

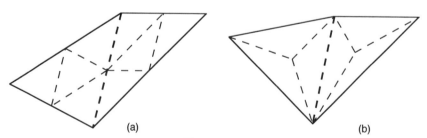

Figure 2.2
Admissible (a) and nonadmissible (b) linear macro-elements for Stenberg's method of verifying the div-stability condition. ___ Finite element boundary, ___ macro-element and element boundary, ___ common boundary of two-macro-elements.

local computation to determine the dimension of the null spaces (2.31). This condition need be checked only once for each type of macro-element. The additional geometric condition needed in the linear or bilinear case is also easy to verify.

Stenberg's method has been used, e.g., in Silvester and Thatcher [1986], Stenberg [1984 and 1987], and Thatcher and Silvester [1987], to verify the div-stability of some two-dimensional finite element spaces. See Stenberg [1987] for a more general method of verifying the div-stability condition.

2.5. Examples of Unstable Spaces Including the Bilinear-Constant Pair

There are different ways in which arbitrarily chosen finite element spaces may fail to satisfy the div-stability condition. Here we discuss some of these and then give specific examples, focusing on the much studied and much misunderstood bilinear velocity-constant pressure pair.

The most catastrophic type of failure is for (2.2), or equivalently (2.4), to imply that $\mathbf{u}^h = \mathbf{0}$, i.e., the only discretely solenoidal field belonging to $\mathbf{V}_0^h$ is the zero vector. The approximate solution is useless since, of course, $\mathbf{u}^h = \mathbf{0}$ cannot be a good approximate solution of the Navier–Stokes equations. This type of situation can usually be detected by a counting argument, i.e., the discrete divergence matrix $b(\mathbf{v}_k, q_j)$, $j = 1, ..., J$, and $k = 1, ..., K$, appearing in (2.4) has more independent rows than columns.

Less catastrophic is the situation wherein for one or a few, but not all, $q^h \in S_0^h$ we have that $b(\mathbf{v}^h, q^h) = 0$ for all $\mathbf{v}^h \in \mathbf{V}_0^h$ so that $\gamma = 0$ in (2.10). This kind of failure of the div-stability condition is also usually easy to detect since it results, in practice, in the discrete divergence matrix being rank deficient. Furthermore, if this type of pressure mode is the sole reason for the invalidity of (2.10), one may often still obtain, through a filtering process, useful approximations.

A more subtle failure of the div-stability condition is the case where, for at least some $q^h \in S_0^h$,

$$C_1 h \|q^h\|_0 \leq \sup_{\mathbf{0} \neq \mathbf{v}^h \in \mathbf{V}_0^h} \frac{b(\mathbf{v}^h, q^h)}{|\mathbf{v}^h|_1} \leq C_2 h \|q^h\|_0 \tag{2.32}$$

for some constants C_1 and C_2 independent of h. In this case $\gamma = O(h)$ where γ is the constant appearing in (2.10). In practice this may result in a loss of accuracy, especially for the pressure approximations. Such instabilities are harder to detect, because, of course, computations are carried out using a finite value of h. In particular, no problems such as those caused by rank deficient approximations to the continuity equation are encountered. This type of situation points out the dangers of calculating on only one grid and of not at least performing serious mesh refinement studies. It also points out the usefulness of rigorous results concerning the stability, or lack thereof, of finite element spaces.

An Unstable Linear-Constant Pair.   An example of the first and most catastrophic instability is the following seemingly natural choice for the velocity and pressure finite element spaces. Let Ω be a square that is triangulated as indicated in Figure 2.3. For the velocity approximations, we choose piecewise linear functions with respect to the given triangulation that are continuous over $\bar{\Omega}$ and that vanish on Γ. For the discrete pressures, we choose piecewise constant functions with respect to the same triangulation and that have zero mean over Ω. Clearly, $\mathbf{V}_0^h \subset \mathbf{H}_0^1(\Omega)$ and $S_0^h \subset L_0^2(\Omega)$. For this choice the only discrete velocity field $\mathbf{u}^h \in \mathbf{V}_0^h$ satisfying the discrete incompressibility

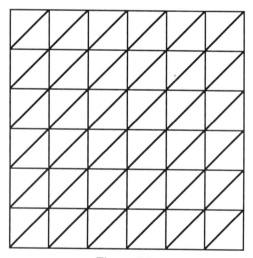

Figure 2.3
A subdivision of a square into triangular finite elements.

constraint (2.2) is $\mathbf{u}^h = \mathbf{0}$, i.e., the only discretely solenoidal velocity field is the zero vector! One easily sees that if there are N cells to a side, the number of equations in (2.4) is $J = \dim(S_0^h) = 2N^2 - 1$, which is greater than the number of unknowns $K = \dim(\mathbf{V}_0^h) = 2(N - 1)^2$. The independence of the equations can be easily established.

In the above example, we see that the discrete incompressibility condition (2.2) imposes too many constraints relative to the available velocity degrees of freedom. In fact, $\dim(S_0^h) > \dim(\mathbf{V}_0^h)$. In order to remedy the situation, one must at least increase the dimension of $\mathbf{V}_0^h$ relative to that of S_0^h.

The Bilinear-Constant Element Pair. We next consider the bilinear velocity-constant pressure pair, which is often referred to as the $Q_1 - P_0$ element pair. Again consider the case of Ω being a square and consider the "triangulation" of Figure 2.4. We now choose $\mathbf{V}_0^h$ to consist of piecewise bilinear functions with respect to this triangulation that are continuous over $\bar{\Omega}$ and that vanish on Γ. For S_0^h we choose piecewise constant functions over the same triangulation that have zero mean over Ω. Once again the inclusions $\mathbf{V}_0^h \subset \mathbf{H}_0^1(\Omega)$ and $S_0^h \subset L_0^2(\Omega)$ hold. The simple counting argument used for the first example does not yield any definitive information since $\dim(\mathbf{V}_0^h) = 2(N - 1)^2$, the same as before, while now $\dim(S_0^h) = N^2 - 1$.

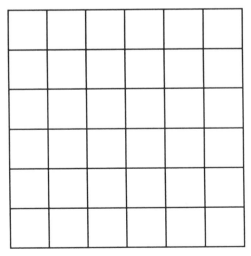

Figure 2.4
A subdivision of a square into rectangular finite elements.

It is well known, e.g., see Brooks and Hughes [1982], Fortin [1977], Gunzburger, Nicolaides, and Peterson [1982], Johnson and Pitkaranta [1982] and Sani *et al.* [1981], that this bilinear-constant element pair exhibits the disastrous "checkerboard" mode, i.e., for the particular discrete pressure field $q^h \in S_0^h$ that is +1 in the "red squares" and −1 in the "black squares," we have that $b(\mathbf{v}^h, q^h) = 0$ for all $\mathbf{v}^h \in \mathbf{V}_0^h$. This is an example of the second type of instability discussed above. The single "bad" pressure mode can be easily filtered out, and therefore some have suggested that once this mode is taken care of, the bilinear-constant element pair can be safely used.

However, this is not the whole story for the bilinear-constant element pair. Boland and Nicolaides [1984] have shown that there exist other pressure modes for which (2.32) is satisfied. The left-hand inequality of (2.32) was previously known (Johnson and Pitkaranta [1982]), at least in the different context of penalty methods. Of course, the left inequality does not imply the right, and certainly doesn't imply that for those modes the stability constant $\gamma = O(h)$. However, Boland and Nicolaides have shown that this is indeed the case. Moreover, they have shown (Boland and Nicolaides [1985]) that there exist data $\mathbf{f}$ for which the pressure approximations do not converge and that it is also possible to set up problems for which the velocity approximations do not converge as well. At the least, since the constants in the error estimates (2.16)–(2.20) are proportional to $1/\gamma$, there will likely be a loss of accuracy due to these pressure modes.

These conclusions are worth noting, especially in view of the fact that the bilinear-constant element pair, with the checkerboard mode filtered out, has been used on numerous occasions in practical computations. It is probably the case that it is safe to use this element pair to define approximations of the velocity, at least for problems with homogeneous boundary conditions. However, there is considerable danger that pressure approximations are poor unless some further post-processing (smoothing) is done over and above removing the checkerboard mode.

3

Finite Element Spaces

In this chapter we discuss pressure and velocity finite element spaces that have been rigorously shown to satisfy the div-stability condition. There are many such pairs known, especially for two-dimensional problems; therefore we will restrict our attention to pairs that have the greatest practical utility. We will also discuss some means for obtaining useful approximations while circumventing the div-stability condition.

Throughout, $P_k(\mathfrak{D})$ denotes the space of polynomials of degree less than or equal to k with respect to the set $\mathfrak{D} \subset \mathbb{R}^n$, and $\mathbf{P}_k(\mathfrak{D})$ denotes the space of n-vector valued functions each of whose components belongs to $P_k(\mathfrak{D})$. Analogous definitions hold for $Q_k(\mathfrak{D})$ and $\mathbf{Q}_k(\mathfrak{D})$ in the case of functions that are polynomials of degree less than or equal to k in each of the coordinate directions, e.g., $Q_1(\mathfrak{D})$, for $\mathfrak{D} \subset \mathbb{R}^2$, denotes piecewise bilinear functions with respect to the set $\mathfrak{D}$. Likewise, we define the spaces $C^k(\mathfrak{D})$ and $\mathbf{C}^k(\mathfrak{D})$ of k-times continuously differentiable functions with respect to the set $\mathfrak{D}$.

For the most part, the results below hold for polygonal domains in $\mathbb{R}^2$ or polyhedral domains in $\mathbb{R}^3$. Through the use of, e.g., isoparametric elements, most of the results will also hold for domains with curved boundaries, provided the latter satisfy appropriate smoothness criteria.

Furthermore, we assume that all subdivisions of Ω into finite elements that are employed below satisfy the standard regularity conditions. For details concerning these issues, one may consult, e.g., Ciarlet [1978].

We remark that it is usually cumbersome to impose the global zero mean over the Ω constraint directly on the discrete pressure spaces. What is usually done is to ignore this constraint when computing a discrete pressure and then impose it *a posteriori* by subtracting, from the computed pressure, its mean. Thus below, when discussing the basis functions for the discrete pressure spaces, we do not impose the zero mean constraint.

3.1. Piecewise Linear and Bilinear Velocity Fields

We begin with some examples involving piecewise linear or bilinear velocity fields with respect to a subdivision of Ω into triangles or rectangles, respectively. In all cases the discrete velocity fields are continuous over $\bar{\Omega}$. In combination with these types of velocity finite element spaces, we will consider both discontinuous piecewise constant and continuous, over $\bar{\Omega}$, piecewise linear pressure fields. Every element pair listed satisfies the div-stability condition (2.10). Moreover, provided the solution $\mathbf{u}, p$ of (1.9)–(1.10) has the indicated smoothness, the following error estimates for the discrete solution $\mathbf{u}^h$, p^h of (2.1)–(2.2) hold uniformly in h:

$$
\left.
\begin{aligned}
|\mathbf{u} - \mathbf{u}^h|_1 &= O(h) \\
\|\mathbf{u} - \mathbf{u}^h\|_0 &= O(h^2) \\
\|p - p^h\|_0 &= O(h)
\end{aligned}
\right\} \text{ whenever }
\left\{
\begin{aligned}
\mathbf{u} &\in \mathbf{H}^2(\Omega) \cap \mathbf{H}_0^1(\Omega) \\
&\text{and} \\
p &\in H^1(\Omega) \cap L_0^2(\Omega)
\end{aligned}
\right\}. \tag{3.1}
$$

Thus, these elements yield, with respect to $L^2(\Omega)$-norms, first-order accurate pressure approximations and second-order accurate velocity approximations.

Piecewise Constant Pressures I. For the linear-constant element pair mentioned in Section 2.5, the discrete continuity equation overconstrained the approximate velocity field. However, by employing different grids for the pressure and velocity fields, the linear-constant element pair may be made stable. For example, consider a given triangulation $\mathfrak{I}_h$ of a polygonal domain $\bar{\Omega}$ into triangles. Then divide

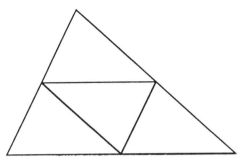

Figure 3.1
A pressure triangle and its four associated velocity triangles for the first piecewise constant pressure element.

each triangle in $\mathfrak{I}_h$ into four triangles by joining the midsides, thus defining a refined triangulation $\mathfrak{I}_{h/2}$. An example is provided in Figure 3.1. Now define

$$V_0^h = \{v : v \in P_1(\Delta), \, \Delta \in \mathfrak{I}_{h/2}; \quad v \in C^0(\bar{\Omega}); \quad v = 0 \text{ on } \Gamma\}$$

$$S_0^h = \left\{q : q \in P_0(\Delta), \, \Delta \in \mathfrak{I}_h; \quad \int_\Omega q \, d\Omega = 0\right\}$$

(3.2)

so that the pressure is sought among piecewise constant polynomials with respect to the triangulation $\mathfrak{I}_h$ while the velocity is sought among continuous piecewise linear polynomials with respect to the finer triangulation $\mathfrak{I}_{h/2}$. The velocity degrees of freedom are simply function values at the interior vertices (with respect to Ω) of the triangles in $\mathfrak{I}_{h/2}$. There is one pressure degree of freedom associated with each triangle in $\mathfrak{I}_h$. The pair of finite element spaces defined by (3.2) are known to satisfy the div-stability condition (2.10) and thus yield optimally accurate approximations satisfying (3.1).

Piecewise Constant Pressures II. For the unstable linear-constant element pair of Section 2.5, there was one velocity element for each pressure element; for the stable linear-constant element pair (3.2), there are four velocity triangles for each pressure triangle. Stable linear-constant element pairs may be defined wherein the ratio of discrete velocities to pressures is not so high. For example, let the velocity space V_0^h be as in (3.2); now define the pressure space S_0^h through the following choice of basis. For each triangle of $\mathfrak{I}_h$ we define three basis functions, namely, piecewise constants that are unity in

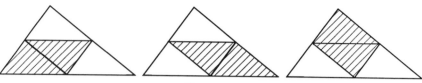

Figure 3.2

Support of the pressure basis functions associated with four velocity triangles for the second piecewise constant pressure element.

the shaded areas in Figure 3.2 and zero in the unshaded areas. Of course, outside the particular triangle of $\mathfrak{I}_h$, the basis functions vanish as well. This pressure space consists of three out of the four possible piecewise constants associated with the four triangles in $\mathfrak{I}_{h/2}$ contained within a single triangle in $\mathfrak{I}_h$. Moreover, there are essentially three times as many pressure degrees of freedom for this choice of S_0^h as there are for the choice made in (3.2). However, this element pair is also stable, i.e., satisfies the div-stability condition (2.10) and the error estimates (3.1).

We remark that it seems that in order to produce div-stable finite element pairs, it is necessary to gain some control of the velocity, or at least the normal velocity, i.e., the mass flux, along the edges of the triangles defining the pressure space. For example, the choice of finite element spaces depicted in Figure 3.3 can be easily shown to be *not* div-stable. Note that the velocity triangulation is derived from the pressure triangulation by dividing each pressure triangle into three triangles by joining the vertices to the centroid. Furthermore, note that there are no velocity nodes present along the edges of a pressure triangle.

Piecewise Linear Pressures.   One may also couple a piecewise linear velocity element with a piecewise linear pressure element and

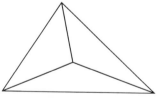

Figure 3.3

A pressure triangle and its three associated velocity triangles for a linear velocity–constant pressure element combination that is not div-stable.

still satisfy the div-stability condition (2.10) and the estimates (3.1). Such a pair was introduced in Bercovier and Pironneau [1979], was analyzed there and in Verfurth [1984a], and is given by

$$S_0^h = \left\{ q : q \in P_1(\Delta),\ \Delta \in \mathfrak{I}_h;\quad q \in C^0(\bar{\Omega});\quad \int_\Omega q\, d\Omega = 0 \right\} \tag{3.3}$$

$$V_0^h = \{ \mathbf{v} : \mathbf{v} \in \mathbf{P}_1(\Delta),\ \Delta \in \mathfrak{I}_{h/2};\quad \mathbf{v} \in \mathbf{C}^0(\bar{\Omega});\quad \mathbf{v} = 0 \text{ on } \Gamma \}.$$

The pressure degrees of freedom are now function values at the vertices of the triangles in $\mathfrak{I}_h$.

Due to the coupling between the pressure and velocity errors in the estimates (2.16) and (2.17), one seemingly cannot take advantage of the better approximating ability of the linear pressure space. Thus, insofar as the rates of convergence, this linear–linear element pair is no better than the stable linear–constant element pairs. However, in practical calculations we have found this to be the best element combination involving linear velocity fields, better in the sense of giving more accurate results for useful values of h. Experimental evidence indicates that for the linear–constant element pairs the second terms, i.e., the ones involving the ability to approximate in the pressure space, appearing in the right-hand side of the error estimates (2.16) and (2.17) are large compared to the first terms, i.e., the ones involving the ability to approximate in the velocity space. This is due, in part, to the fact that the constants C_2 and C_4 are larger than C_1 and C_3, respectively, and also to the fact that the pressure triangulation is coarser than that for the velocity. The extra power of h resulting from the use of piecewise linear pressures mitigates this disparity in the size of the two terms appearing in the error estimates.

Another advantage of this linear–linear element pair is that it usually results in fewer unknowns, for the same grid, than do the linear–constant pairs. For example, suppose the pressure triangulation $\mathfrak{I}_h$ is given by Figure 2.3 with N intervals on each side. Thus there are $2N^2$ triangles in $\mathfrak{I}_h$ and the element pair (3.2) has $2N^2 - 1$ pressure unknowns; on the other hand, the number of nodes in this triangulation is only $(N + 1)^2$, and thus the piecewise linear pressure space of (3.3) has only $(N + 1)^2 - 1$ degrees of freedom. Both element pairs have $2(2N - 1)^2$ velocity unknowns so that the linear–linear element pair (3.3) has roughly N^2 less degrees of freedom, for the same grid, as does the linear–constant element pair (3.2).

Piecewise Bilinear Velocity Fields. Entirely analogous to the triangular elements described above, we have the following elements involving bilinear velocity fields with respect to rectangular elements. More general quadrilateral elements may be found from these through, e.g., isoparametric mappings.

We start with a subdivision $\mathcal{Q}_h$ of $\bar{\Omega}$ into rectangles, or more generally, quadrilaterals. Subsequently we divide each rectangle into four smaller rectangles by joining the opposite midsides, thus creating another subdivision $\mathcal{Q}_{h/2}$ of $\bar{\Omega}$ into rectangles. See Figure 3.4. In all three velocity-pressure element pairs about to be described, we choose the approximating velocity space to consist of piecewise bilinear functions with respect to the subdivision $\mathcal{Q}_{h/2}$, which are continuous over $\bar{\Omega}$ and which vanish on Γ, i.e.,

$$\mathbf{V}_0^h = \{\mathbf{v} : \mathbf{v} \in \mathbf{Q}_1(\square), \square \in \mathcal{Q}_{h/2}; \quad \mathbf{v} \in \mathbf{C}^0(\bar{\Omega}); \quad \mathbf{v} = \mathbf{0} \text{ on } \Gamma\}. \quad (3.4)$$

The velocity degrees of freedom are simply function values at the interior vertices of the rectangles in $\mathcal{Q}_{h/2}$.

For the first pressure space we choose piecewise constants with respect to the larger quadrilaterals of the subdivision $\mathcal{Q}_h$ and that have zero mean over Ω, i.e.,

$$S_0^h = \left\{ q : q \in Q_0(\square), \square \in \mathcal{Q}_h; \quad \int_\Omega q \, d\Omega = 0 \right\}.$$

There is one pressure degree of freedom associated with each rectangle in $\mathcal{Q}_h$. As is indicated in Figure 3.5, for the second pressure space we

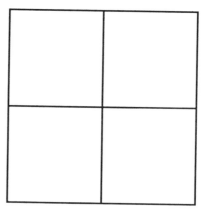

Figure 3.4
A pressure quadrilateral and its four associated velocity quadrilaterals.

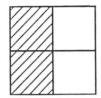 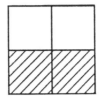 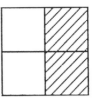

Figure 3.5

Support of the three pressure basis functions associated with four velocity quadrilaterals.

choose three out of the four possible piecewise constants associated with the rectangles belonging to $Q_{h/2}$ and which have zero mean over Ω. Finally, the third pressure space consists of piecewise bilinear functions with respect to the subdivision Q_h that are continuous over $\bar{\Omega}$ and have zero mean over Ω, i.e.,

$$S_0^h = \left\{ q : q \in Q_1(\square), \; \square \in Q_h; \quad q \in C^0(\bar{\Omega}); \quad \int_\Omega q \, d\Omega = 0 \right\}. \quad (3.5)$$

Now the pressure degrees of freedom are function values at the vertices of the rectangles in Q_h.

The three velocity–pressure elements just described satisfy the div-stability condition (2.10) and the error estimates (3.1). Similar to the case for triangles and for the same reasons, the preferred element pair involving bilinear velocities is (3.4) coupled with (3.5), i.e., the bilinear velocity-bilinear pressure pair.

3.2. The Taylor–Hood Element Pair

We next turn to quadratic and biquadratic approximate velocity fields. Suppose we have a triangulation $\mathfrak{I}_h$ of $\bar{\Omega}$. Then the Taylor–Hood element pair (see Taylor and Hood [1973]) is defined by

$$V_0^h = \{ \mathbf{v} : \mathbf{v} \in P_2(\Delta), \; \Delta \in \mathfrak{I}_h; \quad \mathbf{v} \in C^0(\bar{\Omega}); \quad \mathbf{v} = \mathbf{0} \text{ on } \Gamma \}$$

$$S_0^h = \left\{ q : q \in P_1(\Delta), \; \Delta \in \mathfrak{I}_h; \quad q \in C^0(\bar{\Omega}); \quad \int_\Omega q \, d\Omega = 0 \right\}. \quad (3.6)$$

Note that we are now basing V_0^h and S_0^h on the same grid but on different degree polynomials, in contrast to (3.3), which uses the same degree polynomials but different grids. The velocity degrees of freedom are function values at the interior vertices and interior midsides of the triangles in $\mathfrak{I}_h$; the pressure degrees of freedom are function values at the vertices of these triangles. The element pair (3.6) satisfies the div-stability condition (2.10). (Incidentally, although widely believed, it is not necessary to triangulate into corners in order to achieve optimal accuracy; this was established in Silvester and Thatcher [1986]. The triangulation of Figure 2.3, unlike the slightly different triangulation of Figure 3.6, fails to triangulate into the upper-left and lower-right corners.) Furthermore, if the solution $\mathbf{u}$, p of (1.9)–(1.10) has the indicated smoothness, then the following error estimates hold uniformly in h:

$$\left.\begin{aligned} |\mathbf{u} - \mathbf{u}^h|_1 &= O(h^{m-1}) \\ \|\mathbf{u} - \mathbf{u}^h\|_0 &= O(h^m) \\ \|p - p^h\|_0 &= O(h^{m-1}) \end{aligned}\right\} \text{ whenever } \left\{\begin{aligned} \mathbf{u} &\in \mathbf{H}^m(\Omega) \cap \mathbf{H}_0^1(\Omega) \\ &\text{and} \\ p &\in H^{m-1}(\Omega) \cap L_0^2(\Omega) \end{aligned}\right\},$$

$$m = 2 \text{ or } 3. \quad (3.7)$$

These results have been established by many authors, including Bercovier and Pironneau [1979], Boland and Nicolaides [1983], Silvester and Thatcher [1986], and Verfurth [1984a]. We see from (3.7) that if $\mathbf{u} \in \mathbf{H}^3(\Omega) \cap \mathbf{H}_0^1(\Omega)$ and $p \in H^2(\Omega) \cap L_0^2(\Omega)$ then, in L^2-norms, we have third-order accurate velocity approximations and second-order accurate pressure approximations. This is an improvement over any of the elements involving linear velocities.

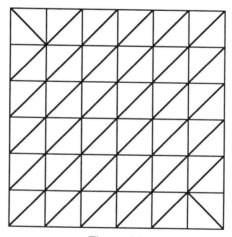

Figure 3.6
Triangulating into corners.

One should note that the number of degrees of freedom, both of velocity and pressure type, associated with the use of (3.6) is identical to that associated with the use of (3.3), the most efficient linear velocity element. In fact, the structure of the discrete system resulting from a Taylor-Hood discretization is in every way identical to that resulting from the use of (3.3). Therefore, the solution times for the Taylor-Hood and the linear-linear discrete systems are roughly the same if one uses the same pressure triangulation in both cases. Of course, the Taylor-Hood element pair will yield better accuracy than the linear-linear pair, provided the exact solution is sufficiently smooth.

On the other hand, on the same grid, the assembly costs associated with the Taylor-Hood element pair will in general be higher since one needs to use higher-order quadrature rules to integrate higher degree polynomial integrands. See Section 4.4 below. For many solvers, the assembly time is overwhelmed by the solution time; therefore the increased assembly cost associated with (3.6) is not a serious drawback. Of course, this is further mitigated by the fact that for the same accuracy, one may use a coarser grid with (3.6) than with (3.3).

Summarizing, provided the exact solution is sufficiently smooth, the Taylor-Hood element pair, when compared to any of the linear velocity elements, yields better accuracy for essentially the same work, or alternately, will yield a desired level of accuracy for less cost.

For rectangles or quadrilaterals we have the analogous pair

$$\mathbf{V}_0^h = \{\mathbf{v} : \mathbf{v} \in \mathbf{Q}_2(\square), \ \square \in \mathcal{Q}_h; \quad \mathbf{v} \in C^0(\bar{\Omega}); \quad \mathbf{v} = \mathbf{0} \text{ on } \Gamma\}$$

$$S_0^h = \left\{ q : q \in Q_1(\square), \ \square \in \mathcal{Q}_h; \quad q \in C^0(\bar{\Omega}); \quad \int_\Omega q \, d\Omega = 0 \right\}, \tag{3.8}$$

where $\mathcal{Q}_h$ denotes a subdivision of $\bar{\Omega}$ into rectangles. This element pair satisfies the div-stability condition (2.10) and the error estimates (3.7); see Stenberg [1984].

The pressure degrees of freedom for the element pair (3.8) are function values at the vertices of the rectangles in $\mathcal{Q}_h$. The velocity degrees of freedom are function values at the interior vertices, interior midsides, and centroids of those rectangles. A variant of the element pair (3.8) is the "serendipity" element, wherein the velocity degrees of freedom associated with the centroids of the rectangles in $\mathcal{Q}_h$ are deleted. (See Figure 3.7.) In this case the velocity finite element space $\mathbf{V}_0^h$ restricted to a rectangle $\square$ is a strict subspace of $\mathbf{Q}_2(\square)$; however,

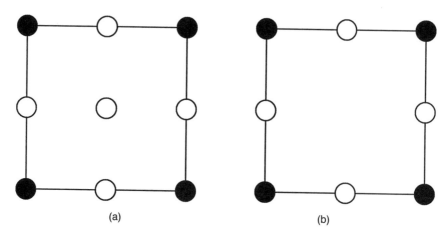

(a)   (b)

Figure 3.7

Two quadrilateral Taylor–Hood element pairs. (a) Biquadratic velocity–bilinear pressure element and (b) biquadratic velocity–bilinear pressure serendipity element. ● Velocity and pressure node, ○ velocity node.

it still contains $\mathbf{P}_2(\square)$, so that its approximation theoretic rate of convergence is the same as that for the velocity space of (3.8). Moreover, this serendipity velocity space, coupled with the pressure space of (3.8), is div-stable (see Stenberg [1984]), and thus yields the same error estimates as does (3.8), i.e., (3.7). On the other hand, the serendipity element requires one less velocity (vector) unknown per rectangle.

One may well ask if further efficiencies may be gained by using higher-order elements, e.g., cubic velocities coupled with quadratic pressures. Here one needs to consider the trade-off between the increased accuracy of higher-order elements and the increased complexity of those elements. As in other settings, e.g., structural mechanics, one generally finds that the optimum seems to be achieved by quadratic elements. Furthermore, it is questionable that in general settings the exact solution of the Navier–Stokes equations is sufficiently smooth to enable the potential better accuracy of higher-order elements. See Section 3.4 for a further discussion.

The Taylor–Hood element pair (3.6), or its quadrilateral counterparts, is a good choice in many settings. However, in some instances, e.g., when there is a large amount of recirculation, it is found that the Taylor–Hood element pair yields unsatisfactory streamline patterns. See Thatcher and Silvester [1987], Tidd, Thatcher, and Kaye [1986], and the discussion in Section 3.6.

3.3. Bubble Elements, the $Q_2 - P_1$ Element Pair, and Other Elements of Interest

Another class of div-stable elements is generally known as *bubble elements* due to the appearance of "bubble"-like functions in the basis sets for the discrete velocity space. These are examples of finite element spaces for which stability is achieved by enlarging the velocity space, which may be compared with the linear elements discussed above, for which stability is achieved by restricting the pressure space. Bubble elements are discussed in, e.g., Boland and Nicolaides [1983], Brezzi and Pitkaranta [1984], and Mansfield [1982]. For the pressure space one chooses a complete and discontinuous piecewise polynomial space with respect to some triangulation $\mathfrak{I}_h$ of $\bar{\Omega}$, i.e., for $k \geq 2$,

$$S_0^h = \left\{ q : q \in P_{k-1}(\Delta), \ \Delta \in \mathfrak{I}_h; \quad \int_\Omega q \, d\Omega = 0 \right\}, \qquad (3.9)$$

and for the velocity space the complete and continuous piecewise polynomial space of one degree higher is augmented with bubble functions in each triangle, i.e.,

$$\mathbf{V}_0^h = \{\mathbf{v} : \mathbf{v} \in [\mathbf{P}_k(\Delta) \oplus \mathbf{B}_k(\Delta)], \ \Delta \in \mathfrak{I}_h; \quad \mathbf{v} \in \mathbf{C}^0(\bar{\Omega}); \quad \mathbf{v} = \mathbf{0} \text{ on } \Gamma\}, \qquad (3.10)$$

where, for any $\Delta \in \mathfrak{I}_h$,

$$\mathbf{B}_k(\Delta) = \{\mathbf{v} \in \mathbf{P}_{k+1}(\Delta) : \mathbf{v} = \lambda_1 \lambda_2 \lambda_3 \mathbf{u}, \quad \mathbf{u} \in \mathbf{P}_{k-2}(\Delta)\}.$$

Here, $\lambda_i(\mathbf{x})$, $i = 1, 2, 3$, denote the barycentric coordinates of the point $\mathbf{x} \in \Delta$ with respect to the vertices of Δ; see Ciarlet [1978]. Thus, for example, if $k = 2$, S_0^h consists of discontinuous piecewise linear polynomials with respect to $\mathfrak{I}_h$, while $\mathbf{V}_0^h$ consists of vector valued functions each of whose components is a continuous piecewise quadratic polynomial augmented in each triangle $\Delta \in \mathfrak{I}_h$ by the cubic bubble function that vanishes on the three edges of Δ. For $k \geq 2$, if $\mathbf{u} \in \mathbf{H}^{k+1}(\Omega) \cap \mathbf{H}_0^1(\Omega)$ and $p \in H^k(\Omega) \cap L_0^2(\Omega)$, then $|\mathbf{u} - \mathbf{u}^h|_1 + \|p - p^h\|_0 = O(h^k)$, and if Ω is convex, $\|\mathbf{u} - \mathbf{u}^h\|_0 = O(h^{k+1})$. Of course, the same accuracy may be achieved using continuous discrete pressure spaces and the corresponding unaugmented velocity spaces. For example, for $k = 2$, one may use the Taylor–Hood pair described above.

Even more general is the following result found in Brezzi and Pitkaranta [1984]. Suppose we have a continuous pressure space, i.e.,

$S_0^h \subset H^1(\Omega) \cap L_0^2(\Omega)$, and a velocity space $\mathbf{V}_0^h$ that includes all continuous piecewise linear functions vanishing on Γ. Or, suppose we have a discontinuous pressure space S_0^h and a velocity space $\mathbf{V}_0^h$ that contains at least all continuous piecewise quadratic functions that vanish on Γ. Then, any such combination of spaces $\mathbf{V}_0^h$ and S_0^h may be stabilized by adding to a basis for $\mathbf{V}_0^h$ the functions $\lambda_1 \lambda_2 \lambda_3 \operatorname{grad}(S_0^h|_\Delta)$ for all $\Delta \in \Im_h$.

An enriched linear velocity bubble element has been analyzed in Arnold, Brezzi, and Fortin [1984]. It is often referred to as the *mini-element*. The velocity finite element space is given by

$$\mathbf{V}_0^h = \{\mathbf{v} : \mathbf{v} \in [P_1(\Delta) \oplus \operatorname{span}(\lambda_1 \lambda_2 \lambda_3)]^2$$

$$\Delta \in \Im_h; \quad \mathbf{v} \in C^0(\bar{\Omega}); \quad \mathbf{v} = \mathbf{0} \text{ on } \Gamma\}$$

and the pressure finite element space by

$$S_0^h = \left\{ q : q \in P_1(\Delta), \Delta \in \Im_h; \quad q \in C^0(\bar{\Omega}); \quad \int_\Omega q \, d\Omega = 0 \right\}.$$

Thus we have continuous piecewise linear velocities enriched by a cubic bubble function in each triangle and continuous piecewise linear pressures. The same triangulation is used for the pressure and velocity. This element pair is div-stable and if $\mathbf{u} \in \mathbf{H}^2(\Omega) \cap \mathbf{H}_0^1(\Omega)$ and $p \in H^1(\Omega) \cap L_0^2(\Omega)$, then $|\mathbf{u} - \mathbf{u}^h|_1 + \|p - p^h\|_0 = O(h)$ and, if Ω is convex, $\|\mathbf{u} - \mathbf{u}^h\|_0 = O(h^2)$.

Another set of elements with discontinuous pressures is discussed in Boland and Nicolaides [1983]. Here we subdivide $\bar{\Omega}$ into quadrilaterals and then subdivide each quadrilateral into two triangles. For S_0^h, one chooses (discontinuous) piecewise polynomials of degree $k - 1$ with respect to the quadrilaterals; for $\mathbf{V}_0^h$, continuous piecewise polynomials of degree k, with respect to the triangles, are used. Then, if $k \geq 2$, such an element pair is div-stable and yields similar error estimates to the bubble elements (3.9)–(3.10).

An important variant of this family doesn't use triangles; one simply chooses $\mathbf{V}_0^h \subset \mathbf{H}_0^1(\Omega)$, restricted to a rectangle $\square$, to be $\mathbf{Q}_k(\square)$-functions. The pressure space remains the same. Again, for $k \geq 2$, the same error estimates hold; see Girault and Raviart [1986] and, for the case $k = 2$, Stenberg [1984]. Note that for both element families the pressure functions are defined over rectangles but are not chosen to be elements of $Q_{k-1}(\square)$; rather, within a rectangle $\square$, the pressure is merely a $P_{k-1}(\square)$-function. The second element pair, with $k = 2$, i.e., the $Q_2 - P_1$

pair with continuous piecewise biquadratic velocities with discontinuous linear pressures defined over rectangles, has been found to be a good element pair in some practical calculations.

3.4. Fourth Degree or Higher Piecewise Polynomial Velocity Spaces

A fundamental change occurs in the relation between velocity and pressure finite element spaces that satisfy the div-stability condition when the polynomials used in the velocity space are of degree higher than three. This fact was demonstrated in Vogelius [1983] and Scott and Vogelius [1985a and 1985b], from which the following material is derived. The importance of these results lies mostly in their implication to the p-version of the finite element method; see Chapter 21.

Given a regular triangulation $\mathfrak{I}_h$ of a connected, polygonal domain $\Omega \subset \mathbb{R}^2$, a discrete velocity space $\mathbf{V}^h$ is defined, for some integer $k \geq 1$, by

$$\mathbf{V}^h = \{\mathbf{v}^h : \mathbf{v}^h|_\Delta \in \mathbf{P}^k(\Delta) \quad \text{for all } \Delta \in \mathfrak{I}_h; \quad \mathbf{v}^h \in \mathbf{C}^{(0)}(\bar{\Omega})\}.$$

We denote, as usual, by $\mathbf{V}_0^h$ the subspace of $\mathbf{V}^h$ consisting of functions that satisfy homogeneous boundary conditions, i.e., $\mathbf{V}_0^h = \mathbf{V}^h \cap \mathbf{H}_0^1(\Omega)$.

In order to precisely define the pressure space, we need to introduce the notions of *singular interior and boundary vertices*. A vertex of a triangle belonging to $\mathfrak{I}_h$ is called a *singular interior vertex* if it does not lie on the boundary Γ of Ω, precisely four edges meet at the point, and these edges lie on two straight lines; see Figure 3.8. A vertex is called

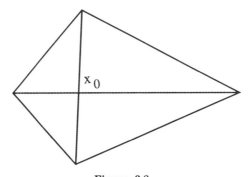

Figure 3.8
A singular interior vertex.

a *singular boundary vertex* if the vertex lies on the boundary and all
the edges of the triangles meeting at the point fall on two straight
lines. See Figure 3.9 for the four possible types of singular boundary
vertices. We denote the sets of all singular interior and boundary
vertices in a given triangulation by S_I and S_B, respectively. For
example, the triangulation depicted in Figure 3.10 contains numerous
singular interior vertices, i.e., at the centers of the rectangles, and no
singular boundary vertices; the one in Figure 2.3 contains no singular
interior vertices but has two singular boundary vertices, i.e., the
upper-left- and lower-right-hand corners; the triangulation of Figure
3.6 contains no singular vertices of either type. For any $\mathbf{x}_s \in S_I$, we

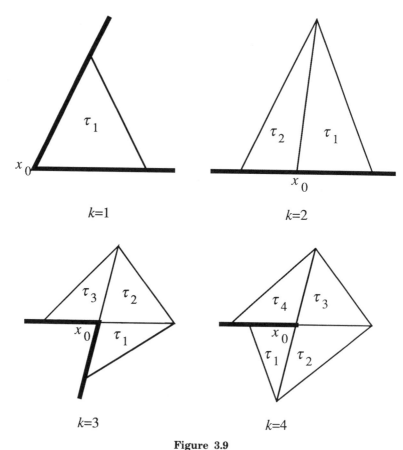

Figure 3.9
Four possible types of singular boundary vertices. ▬▬▬ Edges along boundary Γ,
───── interior edges.

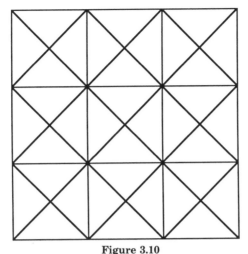

Figure 3.10
A triangulation of a square having numerous singular interior vertices.

denote by Δ_{si}, $i = 1, ..., 4$, the four triangles meeting at $\mathbf{x}_s$, numbered in a counterclockwise manner; see Figure 3.8. Likewise, for any $\mathbf{x}_s \in \mathcal{S}_B$, we denote by Δ_{si}, $i = 1, ..., n_s$, the triangles meeting at $\mathbf{x}_s$, numbered in a counterclockwise manner. Depending on the particular vertex, the value of n_s may be any number from one to four; see Figure 3.9, for which $s = 0$ and $\tau_i = \Delta_{0i}$.

For an integer $k \geq 1$, we introduce the pressure space of discontinuous piecewise polynomials

$$S^h = \left\{ q^h : q^h|_\Delta \in P^{k-1}(\Delta) \quad \text{for all } \Delta \in \mathfrak{I}_h; \right.$$
$$\left. \sum_{i=1}^{4} (-1)^i (q^h|_{\Delta_{si}})(\mathbf{x}_s) = 0 \quad \text{for all } \mathbf{x}_s \in \mathcal{S}_I \right\}$$

and the subspace

$$S_0^h = \left\{ q^h \in S^h : \sum_{i=1}^{n_s} (-1)^i (q^h|_{\Delta_{si}})(\mathbf{x}_s) = 0 \right.$$
$$\left. \text{for all } \mathbf{x}_s \in \mathcal{S}_B; \quad \int_\Omega q^h \, d\Omega = 0 \right\}. \tag{3.11}$$

Thus, for example, the four values of any $q^h \in S^h$ or S_0^h at a singular interior vertex $\mathbf{x}_s$, such as the one of Figure 3.8, must satisfy

$$-q^h|_{\Delta_{s1}}(\mathbf{x}_s) + q^h|_{\Delta_{s2}}(\mathbf{x}_s) - q^h|_{\Delta_{s3}}(\mathbf{x}_s) + q^h|_{\Delta_{s4}}(\mathbf{x}_s) = 0.$$

The first result is that, for any integer $k \geq 1$,

$$\text{div}(\mathbf{V}^h) \subset S^h \quad \text{and} \quad \text{div}(\mathbf{V}_0^h) \subset S_0^h,$$

where, for example, the first of these implies that the divergence of any element of $\mathbf{V}^h$ belongs to S^h. Moreover, for any $k \geq 4$,

$$\text{div}(\mathbf{V}^h) = S^h \quad \text{and} \quad \text{div}(\mathbf{V}_0^h) = S_0^h,$$

so that now, for example, not only does the divergence of any element of $\mathbf{V}^h$ belong to S^h, but also *all* elements of S^h are the divergence of some element of $\mathbf{V}^h$.

Now suppose that nonsingular vertices are *not* nearly singular vertices in the following precise sense. Denote by $\mathbf{x}_0$ a nonsingular vertex and by θ_i, $i = 1, \ldots, n_0$, the angles of the triangles Δ_{0i}, $i = 1, \ldots, n_0$, meeting at $\mathbf{x}_0$. Let

$$R(\mathbf{x}_0) = \max\{|\theta_i + \theta_j - \pi| : 1 \leq i, j \leq n_0 \quad \text{and} \quad i - j = 1 \bmod n_0\}$$

for a nonsingular interior vertex and

$$R(\mathbf{x}_0) = \max\{|\theta_i + \theta_j - \pi| : 1 \leq i, j \leq n_0\}$$

for a nonsingular boundary vertex. For example, in the configuration of Figure 3.11, we have that $R(\mathbf{x}_0) = \delta$. Also let

$$R(\mathfrak{I}_h) = \min\{R(\mathbf{x}_0) : \mathbf{x}_0 \text{ is a nonsingular interior vertex of } \mathfrak{I}_h\}$$

and

$$R_0(\mathfrak{I}_h) = \min\{R(\mathbf{x}_0) : \mathbf{x}_0 \text{ is a nonsingular vertex of } \mathfrak{I}_h\}.$$

Thus, for example, if $R_0(\mathfrak{I}_h) \geq \delta > 0$, where δ may be chosen independent of h, we have that as we refine the triangulation the nonsingular vertices remain uniformly nonsingular.

Armed with these measures of the nonsingularity of nonsingular vertices, one can proceed to obtain the following results. Suppose that $k \geq 4$ and that there exists a positive constant δ, independent of h, such that

$$R(\mathfrak{I}_h) \geq \delta \quad \text{or} \quad R_0(\mathfrak{I}_h) \geq \delta. \tag{3.12}$$

Then,

$$\text{div}(\mathbf{V}^h) = S^h \quad \text{or} \quad \text{div}(\mathbf{V}_0^h) = S_0^h$$

and, given any $q^h \in S^h$ (or $q^h \in S_0^h$), there exists a $\mathbf{v}^h \in \mathbf{V}^h$ (or a $\mathbf{v}^h \in \mathbf{V}_0^h$) such that

$$\text{div } \mathbf{v}^h = q^h \quad \text{and} \quad \gamma \|\mathbf{v}^h\|_1 \leq \|q^h\|_0, \tag{3.13}$$

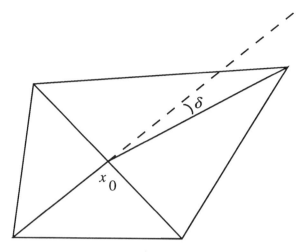

Figure 3.11
A "nearly" singular interior vertex.

where the positive constant γ may be chosen independent of h and of q^h. Note that (3.12) does not rule out the use of singular vertices; it only asks that nonsingular vertices not be close to singular.

One immediate consequence of (3.13) is that if we choose the velocity and pressure spaces to be $\mathbf{V}_0^h$ and S_0^h, respectively and if the triangulation satisfies (3.12), then the spaces $\mathbf{V}_0^h$ satisfy the div-stability condition. To see this, given a $q^h \in S_0^h$, one merely chooses a $\mathbf{v}^h \in \mathbf{V}_0^h$ satisfying (3.13).

Thus, summarizing the results of Scott and Vogelius [1985a], we have the following div-stable choices for the velocity and pressure finite element spaces. Suppose we have a regular triangulation of the polygonal domain Ω that also satisfies (3.12). First, for the velocity, we use piecewise polynomials with respect to the triangulation that, in each triangle, are polynomials of degree less than or equal to k and that are continuous over Ω. We also constrain these polynomials to vanish on the boundary Γ. For the pressure space we choose discontinuous piecewise polynomials with respect to the same triangulation that are of degree $k - 1$ in each triangle. In addition, we constrain these polynomials to satisfy (3.11), which in addition to one global constraint, constitutes one constraint at each singular vertex. Then, if $k \geq 4$, this combination of elements is div-stable, with a stability constant γ independent of h.

Another obvious consequence of (3.13) is that for $k \geq 4$, $\mathbf{Z}^h \subset \mathbf{Z}$, i.e., discretely divergence free functions are actually divergence free. It is also possible to produce, through a rather complicated process, a local basis for $\mathbf{Z}^h$. As we shall see in Section 7.1, this can be a useful thing to have when one solves the discrete equations. There are two keys to producing such local bases. The first is that there is known (see Morgan and Scott [1975] and Scott and Vogelius [1985a]) local bases for "streamfunction" spaces of scalar valued piecewise polynomials of degree $k + 1$ with $k \geq 4$ that are continuously differentiable over Ω and whose values and normal derivative values vanish on the boundary Γ. The second is that, provided Ω is simply connected, the curl operator maps this space into $\mathbf{Z}^h$. Thus, to find a local basis for $\mathbf{Z}^h$, one merely has to take the curl of a local basis for a streamfunction space.

All the results listed above also hold for the case where no boundary conditions on the velocity are imposed.

The accuracy of velocity approximations using these elements is easy to deduce. Since $\mathbf{Z}^h \subset \mathbf{Z}$, the velocity error depends only on the ability to approximate in the velocity space $\mathbf{V}_0^h$. Since in this space we are using complete polynomials of degree k, we have velocity errors, measured in the $\mathbf{H}^1(\Omega)$-norm of $O(h^k)$, *provided* that the exact solution $\mathbf{u} \in \mathbf{H}^{k+1}(\Omega)$. Since here $k \geq 4$, this last condition is not likely to occur in practice. The error in the pressure is more problematical due to the constraints imposed at singular vertices. However, if there are no singular vertices, then we are using complete polynomial spaces for the pressure, and thus, in the $L^2(\Omega)$-norm, the error in the pressure approximation will also be of $O(h^k)$, provided that the exact solution $p \in H^k(\Omega)$.

The fact that local bases for velocity finite element spaces are deduced from local bases for streamfunction spaces leads one to conclude that one should probably use the latter directly in the streamfunction formulation of viscous incompressible flow; see Part V.

3.5. Circumventing the Div-Stability Condition

One way to stabilize unstable elements is to modify the discrete continuity equation. For example, suppose both $\mathbf{V}_0^h$ and S_0^h consist of continuous piecewise linear polynomials with respect to some triangulation $\mathfrak{I}_h$ of $\bar{\Omega}$. This element pair does not satisfy the div-stability condition (2.10). However, in Brezzi and Pitkaranta [1984] it is shown that if (2.2) is modified to

$$b(\mathbf{u}^h, q^h) - \sum_{\Delta \in \mathfrak{I}_h} h_\Delta^2 \int_\Delta \operatorname{grad} p^h \cdot \operatorname{grad} q^h \, d\Omega = 0 \quad \text{for all } q^h \in S_0$$
$$(3.14)$$

where h_Δ denotes the diameter of Δ, then, for the Stokes problem, the resulting discretization is stable and optimally accurate, i.e., if $\mathbf{u} \in \mathbf{H}^2(\Omega) \cap \mathbf{H}_0^1(\Omega)$ and $p \in H^1(\Omega) \cap L_0^2(\Omega)$, then

$$|\mathbf{u} - \mathbf{u}^h|_1 + \|p - p^h\|_0 = O(h).$$

This modification to the discrete continuity equation is also analyzed in Brezzi and Douglas [1988], where an estimate of the form $\|\mathbf{u} - \mathbf{u}^h\|_0 = O(h^2)$ is also provided.

A more general modification of (2.2) was introduced in Hughes, Franca, and Ballestra [1986] and analyzed there and in Brezzi and Douglas [1988]; the method applies to continuous pressure finite element spaces. For the Stokes problem, instead of (3.14), they replace (2.2) with

$$b(\mathbf{u}^h, q^h) + \alpha \sum_{\Delta \in \mathfrak{I}_h} h_\Delta^2 \int_\Delta (\nu \, \Delta \mathbf{u}^h - \operatorname{grad} p^h + \mathbf{f}) \cdot \operatorname{grad} q^h \, d\Omega = 0$$

$$\text{for all } q^h \in S_0^h, \qquad\qquad\qquad (3.15)$$

where α is a constant. The modification actually proposed in Hughes, Franca, and Ballestra [1986] is somewhat more general than (3.15) and can handle a variety of other boundary conditions; for the present purpose, (3.15) will suffice.

The first thing that should be noted about (3.15) is that solutions $\mathbf{u}$ and p of the Stokes problem

$$-\nu \, \Delta \mathbf{u} + \operatorname{grad} p = \mathbf{f} \text{ in } \Omega, \quad \operatorname{div} \mathbf{u} = 0 \text{ in } \Omega, \quad \text{and } \mathbf{u} = \mathbf{0} \text{ on } \Gamma$$
$$(3.16)$$

satisfy (3.15) exactly. Indeed, in each triangle we add to the incompressibility constraint a multiple of the momentum equation. This multiple is chosen to be proportional to h_Δ^2. By appropriately choosing α the problem (2.1) and (3.15) yields stable approximations for finite element pressure and velocity spaces for which the div-stability condition (2.10) is not satisfied.

Now suppose the solution of the Stokes problem satisfies

$$\mathbf{u} \in \mathbf{H}^{r+1}(\Omega) \cap \mathbf{H}_0^1(\Omega) \quad \text{and} \quad p \in H^r(\Omega) \cap L_0^2(\Omega).$$

Also, suppose we use continuous piecewise polynomials of degree $k \geq r$ with respect to some triangulation of $\bar{\Omega}$ for our velocity approximations and continuous piecewise linear polynomials of degree $\ell \geq r - 1$ with respect to the same triangulation for the pressure approximation. Then, it is shown in Brezzi and Douglas [1988], again in the context of the Stokes problem (3.16), that $|\mathbf{u} - \mathbf{u}^h|_1 + \|p - p^h\|_0 = O(h^r)$ and $\|\mathbf{u} - \mathbf{u}^h\|_0 = O(h^{r+1})$. Nothing stops us from choosing $\ell = k$, i.e., using the same degree polynomials and the same mesh for both the velocity and pressure approximations. However, it should be noted that the choice $\ell = k - 1$ will yield the same accuracy, using lower degree polynomials for the pressure, as does the choice $\ell = k$.

In order to get some indication of how the div-stability condition is circumvented, let us consider (3.15) in the context analyzed in Brezzi and Douglas [1988] and in Hughes, Franca, and Ballestra [1986], namely, the Stokes problem (3.16). In order to complete the definition of the discrete equations, we append to (3.15) the discrete momentum equation

$$a(\mathbf{u}^h, \mathbf{v}^h) + b(\mathbf{v}^h, p^h) = (\mathbf{f}, \mathbf{v}^h) \quad \text{for all } \mathbf{v}^h \in \mathbf{V}_0^h. \tag{3.17}$$

Now consider the bilinear form

$$\mathcal{A}([\mathbf{u}^h, p^h], [\mathbf{v}^h, q^h]) = a(\mathbf{u}^h, \mathbf{v}^h) + b(\mathbf{v}^h, p^h) - b(\mathbf{u}^h, q^h)$$

$$- \alpha \sum_{\Delta \in \mathfrak{I}_h} h_\Delta^2 \int_\Delta (\nu \, \Delta \mathbf{u}^h - \operatorname{grad} p^h) \cdot \operatorname{grad} q^h \, d\Omega$$

defined for $[\mathbf{u}^h, p^h]$ and $[\mathbf{v}^h, q^h]$ belonging to the product space $\mathbf{V}_0^h \times S_0^h$. Note that the problem (3.15) and (3.17) can be written as

$$\mathcal{A}([\mathbf{u}^h, p^h], [\mathbf{v}^h, q^h]) = \mathcal{F}([\mathbf{v}^h, q^h]) \quad \text{for all } [\mathbf{v}^h, q^h] \in \mathbf{V}_0^h \times S_0^h, \tag{3.18}$$

where

$$\mathcal{F}([\mathbf{v}^h, q^h]) = (\mathbf{f}, \mathbf{v}^h) + \alpha \sum_{\Delta \in \mathfrak{I}_h} h_\Delta^2 \int_\Delta \mathbf{f} \cdot \operatorname{grad} q^h \, d\Omega.$$

It is shown in Brezzi and Douglas [1988] that, under mild conditions on the triangulation, there exists a constant C such that

$$\mathcal{A}([\mathbf{u}^h, p^h], [\mathbf{u}^h, p^h]) \geq (\nu - C\alpha)|\mathbf{u}^h|_1^2 + \frac{\alpha}{2} \sum_{\Delta \in \mathfrak{I}_h} h_\Delta^2 \int_\Delta \operatorname{grad} p^h \cdot \operatorname{grad} p^h \, d\Omega$$

for all $[\mathbf{u}^h, p^h] \in \mathbf{V}_0^h \times S_0^h$. This implies that for $\alpha > 0$ sufficiently small the bilinear form $\alpha(\cdot, \cdot)$ is coercive over $\mathbf{V}_0^h \times S_0^h$, i.e., there exists a constant $C > 0$, independent of h, such that

$$\alpha([\mathbf{u}^h, p^h], [\mathbf{u}^h, p^h]) \geq C(|\mathbf{u}^h|_1^2 + \|p^h\|_0^2) \qquad \text{for all } [\mathbf{u}^h, p^h] \in \mathbf{V}_0^h \times S_0^h.$$

Then one easily finds that the problem (3.18) has a unique solution $[\mathbf{u}^h, p^h] \in \mathbf{V}_0^h \times S_0^h$; see, e.g., Ciarlet [1978]. But (3.18) is equivalent to (3.15) and (3.17) so that the latter pair also has a unique solution without requiring that $\mathbf{V}_0^h$ and S_0^h satisfy the div-stability condition (2.10). On the other hand, if $\alpha = 0$, the problem (3.15) and (3.17) has an unmodified discrete continuity equation. Furthermore, if $\alpha = 0$, the form $\alpha(\cdot, \cdot)$ is in general no longer coercive on $\mathbf{V}_0^h \times S_0^h$ and the div-stability condition (2.10) is required for the existence and uniqueness of solutions of (3.18), and therefore of (3.15) and (3.17).

Other modifications of the discrete continuity equation that yield stable approximations of solutions of the Stokes problem without requiring (2.10) to be satisfied are proposed and analyzed in Brezzi and Douglas [1988] and in Hughes and Franca [1987]. The latter treats a variety of boundary conditions and presents an algorithm that, in the context of Stokes flow, has the significant added advantage of yielding discrete equations that are symmetric and positive definite. The algorithms proposed in Brezzi and Douglas [1988], Brezzi and Pitkaranta [1984], Hughes and Franca [1987], and Hughes, Franca, and Ballestra [1986] should also be useful in the context of the Navier–Stokes equations. However, no rigorous analyses of these algorithms in that context exists.

We have seen that in order to satisfy the div-stability condition (2.10), one is led to finite element spaces for the velocity and pressure that either or both use different degree polynomials or different meshes. For example, we have the linear-constant element pairs that use different degree polynomials for the velocity and pressure and different meshes, the linear–linear element pair that uses different meshes, and the Taylor–Hood element pair that uses different degree polynomials. Nowhere, so far, have we encountered an element pair that uses the same degree polynomials and the same mesh for both the pressure and the velocity and is also div-stable. However, the modifications of the incompressibility constraint discussed in this section allow for the use of the same degree piecewise polynomials with respect to the same mesh for both the velocity and pressure approximations. This can effect a substantial simplification in finite element codes, both in setting up the geometry and in the assembly of finite element matrices.

It should be noted that penalty methods, which we examine in Chapter 5, can provide, at least for velocity calculations, another method for circumventing the div-stability condition.

3.6. Divergence Free and Mass Conserving Elements

For some applications it is important that mass be conserved, if not exactly, at least very well. Conserving mass exactly requires that the discrete velocity field $\mathbf{u}^h \in \mathbf{Z}$, i.e., the discrete velocity field is, at least in a weak sense, divergence free. Recall that due to (2.2) $\mathbf{u}^h \in \mathbf{Z}^h$, but that in general, $\mathbf{Z}^h \not\subset \mathbf{Z}$. Thus, in principle, one would like to choose the finite element spaces $\mathbf{V}_0^h$ so that the functions belonging to $\mathbf{V}_0^h$ are at least weakly divergence free, i.e., $\mathbf{V}_0^h \subset \mathbf{Z}$. In addition to conserving mass, such a choice effects a great simplification since the determination of the velocity uncouples from that of the pressure. Indeed, since $\mathbf{V}_0^h \subset \mathbf{Z}$ certainly implies that $\mathbf{Z}^h = \mathbf{V}_0^h$, we then need only solve

$$a(\mathbf{u}^h, \mathbf{v}^h) + c(\mathbf{u}^h, \mathbf{u}^h, \mathbf{v}^h) = (\mathbf{f}, \mathbf{v}^h) \quad \text{for all } \mathbf{v}^h \in \mathbf{V}_0^h$$

for the discrete velocity field $\mathbf{u}^h$ since in this case the term $b(\mathbf{v}^h, p^h)$ in (2.1) vanishes for any $\mathbf{v}^h \in \mathbf{V}_0^h = \mathbf{Z}^h$. Also, since $\mathbf{Z}^h \subset \mathbf{Z}$, in the velocity estimate (2.16) $\Theta = 0$, so that the velocity error depends only on the ability to approximate in $\mathbf{V}_0^h$. In fact, the particular choice of pressure space never enters into the velocity calculation, and as a consequence, the div-stability condition is of no import to that computation.

Unfortunately, although there are known some finite element pairs such that the functions in $\mathbf{V}_0^h$ are at least weakly divergence free, these have proven to be impractical, and we will not consider them here. We do mention that one obvious method of generating divergence free discrete vector fields is to take the curl of a piecewise polynomial field, i.e., of a piecewise polynomial streamfunction. One problem with this approach is that if one wants a conforming velocity field, i.e., $\mathbf{V}_0^h \subset \mathbf{H}_0^1(\Omega)$, then the discrete streamfunction field must be chosen to be continuously differentiable over Ω. In $\mathbb{R}^2$ this necessitates the use of at least quintic streamfunctions over triangles (Scott and Vogelius [1985a], or cubic polynomials over macro-elements, e.g., the

Clough–Toucher element (Ciarlet [1978]). See Section 3.4 for a class of divergence free elements that may be derived by this process. There are also known local bases for continuously differentiable piecewise polynomial spaces of degree five or higher (Morgan and Scott [1975]). By taking the curl of these basis functions, one can construct a local basis for the divergence free subspace of the space of continuous piecewise polynomial velocity fields defined over a triangulation of Ω whenever the degree of the polynomials is greater than or equal to four. Of course, once such a smooth discrete streamfunction is introduced, one would be tempted to abandon the primitive variable formulation in favor of the streamfunction formulation of the equations of viscous incompressible flow. See Part V below. Nonconforming velocity fields can also be generated in this manner.

In order to achieve mass conservation, it is not necessary for $\mathbf{V}_0^h \subset \mathbf{Z}$; indeed, since automatically $\mathbf{u}^h \in \mathbf{Z}^h$, it suffices for $\mathbf{Z}^h \subset \mathbf{Z}$. Thus we do not need the whole discrete velocity space $\mathbf{V}_0^h$ to be divergence free, but rather that $\mathbf{V}_0^h$ and S_0^h be chosen such that discretely divergence free functions are also divergence free, at least in a weak sense. Again, although there are some element pairs known such that $\mathbf{Z}^h \subset \mathbf{Z}$ (Fix, Gunzburger, and Nicolaides [1981], Mercier [1979a and 1979b], Nagtegaal, Parks and Rice [1974], and Scott and Vogelius [1985a and 1985b]) these have not proven to be especially practical.

The next best thing to pointwise mass conservation is elementwise mass conservation, i.e., instead of having $\operatorname{div} \mathbf{u}^h = 0$ almost everywhere, we only have that

$$\int_\Delta \operatorname{div} \mathbf{u}^h \, d\Omega = 0 \quad \text{for all } \Delta \in \mathfrak{I}_h. \tag{3.19}$$

The following variant of the Taylor–Hood element pair is known to be div-stable and to satisfy (3.19), i.e., to conserve mass over elements. For the velocity space, we retain the continuous piecewise quadratic finite element space of the Taylor–Hood element pair, i.e., $\mathbf{V}_0^h$ is still defined by (3.6). However, for the pressure space we now use

$$S_0^h = \Big\{ q : q = q_0 + q_1 \, ; \, q_0 \in P_0(\Delta), \, q_1 \in P_1(\Delta), \, \Delta \in \mathfrak{I}_h \, ;$$

$$q_1 \in C^0(\bar{\Omega}) \, ; \, \int_\Omega q_0 \, d\Omega = 0 \, ; \, \int_\Omega q_1 \, d\Omega = 0 \Big\}.$$

Thus the pressure finite element space is essentially that of the Taylor–Hood element pair augmented by a constant in each triangle. This element pair is shown to be div-stable in Thatcher and Silvester [1987]. Furthermore, in Tidd, Thatcher, and Kaye [1986], this element pair was shown to produce computational results, for a particular practical problem, that were superior to those obtainable through the use of the Taylor–Hood element pair. On the other hand, one should note that, for the same mesh, there are roughly three times as many pressure degrees of freedom for the element pair of Thatcher and Silvester than are required by the Taylor–Hood element pair.

The quadrilateral counterpart of this element pair, i.e., continuous piecewise biquadratic velocities coupled with piecewise continuous bilinear pressures enriched with piecewise constant pressures, has been discussed previously, e.g., Gresho, Lee, and Sani [1980], and, in some specific calculations, has been found to be superior to the Taylor–Hood quadrilateral element pair, i.e., continuous piecewise biquadratic velocities coupled with continuous piecewise bilinear pressures.

There are other elements that have been discussed that satisfy the elementwise mass conservation relation (3.19), or its quadrilateral counterpart,

$$\int_\Omega \mathrm{div}\, \mathbf{u}^h \, d\Omega = 0 \qquad \text{for all } \Box \in \mathcal{Q}_h. \tag{3.20}$$

For example, any element pair that involves *discontinuous* pressures will satisfy (3.19) or (3.20), where in these relations $\mathfrak{J}_h$ and $\mathcal{Q}_h$ refer to pressure triangulations. This is easy to see since in (2.2) we may choose the test function q^h to be unity in a single pressure element and to vanish everywhere else. Thus, for example, the $Q_2 - P_1$ element pair discussed in Section 3.3 produces a velocity field that satisfies (3.20).

3.7. Three-Dimensional Elements

Compared to the two-dimensional setting, there are known much fewer stable element pairs for three-dimensional problems. However, there is great interest in this subject and, therefore, there has been substantial recent progress. Here we mention a few of the known stable three-dimensional elements. Some of these, as well as some other three-dimensional element pairs, are discussed in Cuvelier,

Segal, and van Steenhoven [1986], Girault and Raviart [1986], Nedelec [1986], Pitkaranta [1982], Ruas [1985], and Stenberg [1987].

In the first place, the three-dimensional analogue of the Taylor–Hood element is known to be div-stable; this may be shown by the methods of Verfurth, Boland–Nicolaides, or Stenberg. Specifically, we subdivide $\bar{\Omega}$ into tetrahedrons and use continuous piecewise quadratic polynomials for the velocity and continuous piecewise linear polynomials for the pressure. Similarly, for rectangular parallelpipeds, one may use continuous triquadratic velocities and continuous trilinear pressures. The accuracy of these element pairs are the same as their two-dimensional analogues. See Section 3.2.

Next we consider linear–constant elements. Again, subdivide $\bar{\Omega}$ into tetrahedrons. For the pressure space we choose piecewise constants with respect to this initial subdivision. Now we subdivide each tetrahedron into 12 smaller tetrahedrons by first joining the centroid of the faces to the vertices and then joining the centroid of the large tetrahedron to the vertices and the centroids of the faces. For the velocity space, we choose continuous piecewise linear polynomials with respect to the smaller tetrahedrons.

Another linear–constant element pair is defined by first subdividing $\bar{\Omega}$ into rectangular parallelpipeds. For the pressure space we choose piecewise constants over these subregions. We subdivide each parallelpiped into 24 tetrahedrons by first drawing the two diagonals of each face, then joining the centroid of the parallelpiped to the vertices and to the six intersection points of the face diagonals.

Both these linear–constant element pairs are known to be div-stable and yield the same accuracy results as those for the two-dimensional linear–constant pairs.

Another element pair is defined by again subdividing $\bar{\Omega}$ into rectangular parallelpipeds and again choosing piecewise constants over these subregions for the pressure space. We then divide each parallelpiped into eight smaller ones by drawing the lines joining the face centroids. The velocity space is chosen to be piecewise trilinear polynomials with respect to the smaller parallelpipeds. This element pair is known to be div-stable and to yield the same accuracy as do its two-dimensional counterparts of Section 3.1; see Pitkaranta [1982].

A div-stable linear bubble element is given, with respect to rectangular parallelpipeds, by enriching the continuous trilinear velocity finite element space with six bubble functions associated with the faces of each parallelpiped. Each of these bubble functions vanishes at the eight vertices of the parallelpiped and is unity at the centroid of its associated face. This can easily be accomplished with a quintic polynomial. The degrees of freedom are the velocity components at the vertices and the normal velocity component at the centroid of the faces. The pressure is constant in each element. No rigorous analysis of this element pair exists; however, the velocity and pressure approximations seem to be second- and first-order accurate, respectively.

Other possible element pairs that have not been analyzed are the three-dimensional counterparts of the Thatcher–Silvester element (continuous piecewise quadratic or triquadratic velocities and continuous piecewise linear or bilinear pressures enriched by a constant in each element) and the three-dimensional $Q_2 - P_1$ element pair (continuous triquadratic velocities coupled with discontinuous linear pressures). It is quite likely that these element pairs will yield good results.

3.8. Remarks on the Choice of Finite Element Spaces

In this chapter we have examined a large, although certainly not exhaustive, list of stable element pairs. Naturally, one would like to know which, if any, are "best." It is generally thought that elements that at least yield elementwise mass conservation are superior. This judgement is largely based on the examination of graphical representations of the solutions, e.g., streamline plots. Based on this type of criterion, and on previous discussions concerning what is the optimal degree of the polynomials to use, it seems that the preferred element pairs would be those of Thatcher and Silvester [1987], Gresho, Lee, and Sani [1980], and the $Q_2 - P_1$ pair. See Section 3.3 and 3.6. Also note that elementwise mass conservation essentially precludes the use of continuous pressure elements.

There is a price to be paid for the use of elementwise mass conserving elements. As has already been mentioned, these usually involve many more pressure degrees of freedom than do similar stable element pairs with continuous pressures. Furthermore, although the latter often yield poor visual results, e.g., poor flow visualizations, they usually provide more than adequate quantitative information about many functionals of the solution, e.g., drag, pressure forces, etc. Therefore, in some situations, the Taylor–Hood element pair is a good choice; in applications for which good flow visualizations and good resolution of recirculating phenomena are important, the elementwise mass-conserving element pairs mentioned above are probably preferable. If the pressure field is not of great interest, it is also safe to use the bilinear (or trilinear) constant element pair discussed in Section 2.5.

In addition, one should keep in mind that one can circumvent the div-stability condition by using methods such as those discussed in Section 3.5. At one time these methods would have been received with great enthusiasm and would have enjoyed immediate and widespread adoption. However, the incompressibility condition is not so great an obstacle as it once was, so that the methods of Section 3.5 must provide advantages other than avoiding the div-stability condition. They do allow for the use of the same degree polynomials and the same mesh for the pressure and velocity approximations; these features simplify certain programming tasks. Whether these are sufficient advantages to render this approach superior to using div-stable elements is an issue still to be decided. Surely, intrinsic characteristics of the specific problem one wishes to simulate play an important role in the choice of a discretization technique.

We again mention penalty methods, which will be discussed in Chapter 5, and which provide an attractive alternative to the type of methods we have discussed so far.

4

Alternate Weak Forms, Boundary Conditions, and Numerical Integration

In this chapter we examine some variants of the weak formulation (1.9)–(1.10), principally to show how different boundary conditions may be incorporated into finite element methods for the primitive variable formulation and to determine the effects of the use of numerical integration rules when defining the discrete systems. Before considering boundary conditions, we briefly consider an alternate formulation of the convection term in (1.9).

4.1. Alternate Formulations of the Convection Term

For the purpose of simplifying the analysis of the approximate solution, it can be useful to introduce a slightly different weak formulation wherein the trilinear form $c(\cdot, \cdot, \cdot)$ appearing in (1.9) is replaced by the skew-symmetrized form introduced in Temam [1968 and 1979]

$$\tilde{c}(\mathbf{w}, \mathbf{u}, \mathbf{v}) = \tfrac{1}{2}(c(\mathbf{w}, \mathbf{u}, \mathbf{v}) - c(\mathbf{w}, \mathbf{v}, \mathbf{u})). \tag{4.1}$$

One may easily verify that $\tilde{c}(\mathbf{u}, \mathbf{u}, \mathbf{v}) = c(\mathbf{u}, \mathbf{u}, \mathbf{v})$ whenever div $\mathbf{u} = 0$ in Ω and $\mathbf{u} \cdot \mathbf{n} = 0$ or $\mathbf{v} = \mathbf{0}$ on Γ, where $\mathbf{n}$ denotes the outward normal to Γ. Therefore, due to (1.2)–(1.3), it seems irrelevant whether one uses (1.8) or (4.1) in a weak formulation of the Navier–Stokes equations. From an analysis point of view, the advantage of (4.1) is that $\tilde{c}(\mathbf{w}, \mathbf{u}, \mathbf{v}) = -\tilde{c}(\mathbf{w}, \mathbf{v}, \mathbf{u})$ for any $\mathbf{u}, \mathbf{v}, \mathbf{w} \in \mathbf{H}^1(\Omega)$, while the analogous result for (1.8) holds only when div $\mathbf{w} = 0$ in Ω and one of $\mathbf{u} = \mathbf{0}$, $\mathbf{v} = \mathbf{0}$, or $\mathbf{w} \cdot \mathbf{n} = 0$ holds on Γ.

We emphasize that, insofar as the accuracy of the approximations is concerned, it makes no difference whether one uses (1.8) or (4.1); we merely point out that many of the results concerning finite element approximations of solutions of (1.1)–(1.3) were first obtained through the use of (4.1). On the other hand, any implementation of (4.1) will result in more computational work, during the assembly process, than the analogous implementation of (1.8).

4.2. Inhomogeneous Velocity Boundary Conditions

There are many different ways to treat inhomogeneous velocity boundary conditions. In practice, the overwhelming choice is to use the boundary interpolant. We describe this method for polygonal domains $\Omega \subset \mathbb{R}^2$; entirely analogous ideas may be used in three dimensions and for domains with curved sides, the latter through the aid of, e.g., isoparametric elements.

For given $\mathbf{g} \in \mathbf{H}^{1/2}(\Gamma)$, consider the boundary condition

$$\mathbf{u} = \mathbf{g} \text{ on } \Gamma \text{ with } \int_\Gamma \mathbf{g} \cdot \mathbf{n} \, d\Gamma = 0 \tag{4.2}$$

and the set

$$\mathbf{V}_g = \{\mathbf{u} \in \mathbf{H}^1(\Omega) : \mathbf{u} = \mathbf{g} \text{ on } \Gamma, \quad \mathbf{g} \in \mathbf{H}^{1/2}(\Gamma)\}.$$

Note that $\mathbf{V}_0 = \mathbf{H}_0^1(\Omega)$. The weak formulation that we will discretize is as follows: Seek $\mathbf{u} \in \mathbf{V}_g$ and $p \in L_0^2(\Omega)$ such that (1.9) and (1.10) hold. The test function $\mathbf{v}$ still belongs to $\mathbf{H}_0^1(\Omega)$, i.e., $\mathbf{v} = \mathbf{0}$ on Γ.

In order to pose our discrete problem we choose finite element spaces $\mathbf{V}^h \subset \mathbf{H}^1(\Omega)$ and $S_0^h \subset L_0^2(\Omega)$. We denote by $\mathbf{V}^h|_\Gamma$ the restriction of $\mathbf{V}^h$ to the boundary Γ, i.e., $\mathbf{V}^h|_\Gamma$ consists of functions defined on Γ and that agree with the boundary values of at least one function belonging to $\mathbf{V}^h$. The finite element functions belonging to $\mathbf{V}^h$, being, for example, piecewise polynomials, cannot in general satisfy the boundary condition (4.2); certainly, in general $\mathbf{g} \notin \mathbf{V}^h|_\Gamma$. Therefore we choose an approximation to $\mathbf{g}$, which we denote by $\mathbf{g}^h$, belonging to $\mathbf{V}^h|_\Gamma$. The most common choice for $\mathbf{g}^h$, and the one we consider here, is the interpolant of $\mathbf{g}$ in $\mathbf{V}^h|_\Gamma$.

This choice is trivial to implement, which at least partially accounts for its popularity. For example, suppose $\mathbf{V}^h$ is a Lagrangian finite element space, i.e., one whose degrees of freedom are exclusively function values at points. Let $\{\mathbf{v}_k\}$, $k = 1, \ldots, K$, denote a Lagrangian finite element basis for $\mathbf{V}^h$. Let the first K_I of these basis functions be associated with interior nodes $\mathbf{x}_k$ so that for $k = 1, \ldots, K_I$, $\mathbf{v}_k = \mathbf{0}$ for $\mathbf{x} \in \Gamma$. The remaining basis functions $\{\mathbf{v}_k\}$, $k = K_I + 1, \ldots, K$, are associated with nodes $\mathbf{x}_k$ lying on Γ. (In practical implementations there are more efficient node numbering schemes than the one we are using; however, the latter simplifies the explanations being attempted here.) If $\mathbf{g}^h$ is chosen to be the boundary interpolant of $\mathbf{g}$, then the former may be expressed in the form

$$\mathbf{g}^h(\mathbf{x}) = \sum_{k = K_I + 1}^{K} \tilde{\beta}_k \mathbf{v}_k(\mathbf{x}) \qquad \text{for } \mathbf{x} \in \Gamma, \tag{4.3}$$

where the coefficients $\tilde{\beta}_k$ are merely an appropriate component of $\mathbf{g}$ evaluated at the node $\mathbf{x}_k$. In fact, it is customary to choose the basis function $\mathbf{v}_k(\mathbf{x}) = \phi_m(\mathbf{x})\mathbf{e}_i$ where $\mathbf{e}_i$ is one of the coordinate unit vectors and $\phi_m(\mathbf{x})$ is a scalar valued piecewise polynomial such that $\phi_m(\mathbf{x}_\ell) = \delta_{\ell m}$, $\mathbf{x}_\ell$ here denoting a node of the triangulation. We then have that

$$\mathbf{g}^h(\mathbf{x}) = \sum_{m = 1}^{M} \sum_{i = 1}^{n} \phi_m(\mathbf{x})g_i(\mathbf{x}_m)\mathbf{e}_i = \sum_{m = 1}^{M} \phi_m(\mathbf{x})\mathbf{g}(\mathbf{x}_m) \qquad \text{for } \mathbf{x} \in \Gamma,$$

where $\mathbf{x}_m$, for $m = 1, \ldots, M$, denote the nodes on the boundary. Thus, if in (4.3) $\mathbf{v}_k(\mathbf{x}) = \phi_m(\mathbf{x})\mathbf{e}_i$, we have that $\tilde{\beta}_k = \mathbf{g}(\mathbf{x}_m) \cdot \mathbf{e}_i = g_i(\mathbf{x}_m)$ with $k = K_I + (m - 1)n + i$.

If $\mathbf{u}^h \in \mathbf{V}^h$ satisfies $\mathbf{u}^h(\mathbf{x}) = \mathbf{g}^h(\mathbf{x})$ for $\mathbf{x} \in \Gamma$, we have that

$$\mathbf{u}^h(\mathbf{x}) = \sum_{k = 1}^{K_I} \beta_k \mathbf{v}_k(\mathbf{x}) + \sum_{k = K_I + 1}^{K} \tilde{\beta}_k \mathbf{v}_k(\mathbf{x}), \tag{4.4}$$

where β_k, $k = 1, \ldots, K$, are the unknown coefficients to be determined; the coefficients $\tilde{\beta}_k$ of the basis functions associated with boundary nodes are known. The contribution to $\mathbf{u}^h$ emanating from the second summation of (4.4) becomes part of the data of the discrete system of equations.

Once an approximation $\mathbf{g}^h$ is chosen, one may define the set

$$\mathbf{V}_g^h = \{\mathbf{v} \in \mathbf{V}^h : \mathbf{v} = \mathbf{g}^h \text{ on } \Gamma\}.$$

Note that $\mathbf{V}_0^h$ is the finite element subspace of $\mathbf{H}_0^1(\Omega)$ used in conjunction with the homogeneous boundary condition (1.3); also, clearly, $\mathbf{V}_g^h \subset \mathbf{H}^1(\Omega)$ is not a subset of $\mathbf{V}_g$. Now, the approximate problem may be defined as follows: Seek $\mathbf{u}^h \in \mathbf{V}_g^h$ and $p^h \in S_0^h \subset L_0^2(\Omega)$ such that (2.1)–(2.2) hold for all $\mathbf{v}^h \in \mathbf{V}_0^h$ and $q^h \in S_0^h$, respectively. Again, the test functions $\mathbf{v}^h$ vanish on the boundary Γ.

The whole discussion of the div-stability condition (2.10) carries over intact to the case of the inhomogeneous boundary (4.2); in (2.10) we still use the subspace $\mathbf{V}_0^h$ of finite element velocity fields that vanish on the boundary. Results analogous to those of Section 2.3 can be derived in a fairly straightforward manner, with the exception of some technicalities encountered for the $L^2(\Omega)$-error estimate for the velocity approximation. See Fix, Gunzburger and Peterson [1983], Girault and Raviart [1986], Glowinski [1984], and Gunzburger and Peterson [1983] for details.

In particular, if $\mathbf{g}^h$ is chosen to be the boundary interpolant of $\mathbf{g}$ in $\mathbf{V}^h|_\Gamma$, i.e., (4.3), then, provided $\mathbf{g}$ is sufficiently smooth, all the results, e.g., the error estimates, concerning the finite element spaces discussed in Chapter 3 are essentially still valid for the inhomogeneous velocity boundary condition (4.2). Again, see the references cited in the previous paragraph for details.

There are other possible methods for generating discrete velocity boundary conditions. For example, one could define $\mathbf{g}^h$ to be a suitable projection, in $\mathbf{V}^h|_\Gamma$, of $\mathbf{g}$. See Fix, Gunzburger, and Peterson [1983].

We have not addressed the issue of what happens to the compatibility condition in (4.2) when one discretizes. The first observation to be made is that if one computes using a pressure finite element space S_0^h that is a subspace of $L_0^2(\Omega)$, i.e., all $q^h \in S_0^h$ have zero mean over Ω, then one does not need to require that the discretized data satisfies any compatibility condition. The reason for this is that in this case, and if the div-stability condition is satisfied, (2.2) is solvable for any $\mathbf{g}^h$, i.e., the discrete divergence matrix is of full (row) rank.

However, as has already been mentioned, one usually does not impose the global zero mean constraint on the pressure space when one computes. It is more common to either fix the pressure at a point or to impose the constraint after a solution is obtained. In the former case, one again does not encounter any compatibility problems. In the latter case, one works with a discrete pressure space S^h that is merely a subspace of $L^2(\Omega)$, i.e., S^h contains nonvanishing constant functions. Using such an unconstrained pressure space, one finds a discrete pressure $\hat{p}^h$ that does not have zero mean over Ω. However, the discrete pressure $p^h = \hat{p}^h - \alpha$, where α is the mean value of $\hat{p}^h$, then obviously has zero mean.

If one uses discrete pressure spaces whose elements do not have zero mean over Ω, the discrete divergence equation is solvable only if the discrete boundary data satisfies the single compatibility condition

$$\int_\Gamma \mathbf{g}^h \cdot \mathbf{n} \, d\Gamma = 0.$$

For general choices of $\mathbf{g}^h$, this relation is not valid. However, it is easy to show that

$$\left| \int_\Gamma \mathbf{g}^h \cdot \mathbf{n} \, d\Gamma \right| \leq C \|\mathbf{g} - \mathbf{g}^h\|_0$$

so that if $\mathbf{g}^h$ is a good approximation to $\mathbf{g}$, then the necessary compatibility condition on $\mathbf{g}^h$ is nearly satisfied, i.e., to within discretization error. Thus, if one ignores the compatibility condition on $\mathbf{g}^h$, one introduces an error that is of roughly the same size as the error introduced when one approximates the boundary condition, and one can conclude that it is unnecessary to require the data to strictly adhere to the compatibility condition.

If one still wants the discrete data $\mathbf{g}^h$ to satisfy the compatibility condition, then this can be arranged without too much difficulty; see Glowinski [1984].

4.3. Alternate Boundary Conditions and Formulations of the Viscous Term

In this section we examine how different choices for the viscous term in (1.1) affect the natural boundary conditions of corresponding weak formulations. As a consequence we will see how a variety of useful boundary conditions may be easily implemented by finite element methods. For details concerning the material of this section, as well as for the implementation of other boundary conditions, one may consult Conca [1984]; Engelman, Sani, and Gresho [1982]; Girault [1987 and 1988]; Glowinski [1984]; Gunzburger, Nicolaides, and Liu [1985]; Pironneau [1986]; and Verfurth [1987a].

Due to (1.2), whenever v is constant, the viscous term in (1.1) may be written in the various equivalent forms

$$v \, \Delta \mathbf{u} = \tag{4.5.1}$$

$$\mathrm{div}(v((\mathrm{grad} \ \mathbf{u}) + (\mathrm{grad} \ \mathbf{u})^T)) = \tag{4.5.2}$$

$$- v \, \mathrm{curl}(\mathrm{curl} \ \mathbf{u}) = \tag{4.5.3}$$

$$v(\mathrm{grad}(\mathrm{div} \ \mathbf{u}) - \mathrm{curl}(\mathrm{curl} \ \mathbf{u})). \tag{4.5.4}$$

Although these different realizations are equivalent insofar as the partial differential equations are concerned, we shall see how each can generate a different numerical method.

If for some reason v is not constant or $\mathrm{div} \ \mathbf{u} \neq 0$, then only (4.5.2) may be used. Indeed, (4.5.2) is the form of the viscous term that arises naturally in the derivation of the Navier–Stokes equations from the principle of conservation of linear momentum and the Newton–Poisson constitutive equation; see Landau and Lifshitz [1987] or Serrin [1959]. The other three forms (4.5.1), (4.5.3), and (4.5.4) are derived from (4.5.2) with the aid of (1.2) and the assumption that $v = $ constant. In (1.1) we have used (4.5.1) only because this is the most popular choice in the literature; all of the results obtained so far hold equally well if one chooses (4.5.2) instead. As will be seen from the discussion below, (4.5.2) is, in general, to be preferred to (4.5.1).

Denote two segments of the boundary Γ by Γ_n and Γ_τ. These segments may be empty, are not necessarily disjoint, and, in fact, may be equal. See Figure 4.1 for an illustration. Now, for fixed, given functions g_n and $\mathbf{g}_\tau$, define the set

$$\mathbf{V}_g = \{\mathbf{v} \in \mathbf{H}^1(\Omega) : \mathbf{v} \cdot \mathbf{n} = g_n \text{ on } \Gamma_n; \quad \mathbf{n} \times \mathbf{v} \times \mathbf{n} = \mathbf{g}_\tau \text{ on } \Gamma_\tau\}$$

and the spaces

$$\mathbf{V}_0 = \{\mathbf{v} \in \mathbf{H}^1(\Omega) : \mathbf{v} \cdot \mathbf{n} = 0 \text{ on } \Gamma_n; \quad \mathbf{v} \times \mathbf{n} = \mathbf{0} \text{ on } \Gamma_\tau\}$$

and

$$S = L_0^2(\Omega) \text{ if } \Gamma_n = \Gamma, \quad S = L^2(\Omega) \text{ otherwise},$$

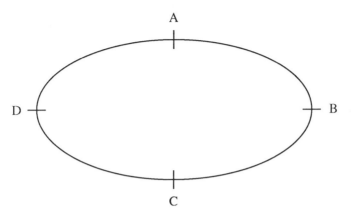

Figure 4.1
An illustration of various boundary segments. $ABC = \Gamma_n$, $BCD = \Gamma_\tau$, $BC = \Gamma_n \cap \Gamma_\tau$, $AD = \Gamma/\Gamma_n \cup \Gamma_\tau$.

where $\mathbf{v} \cdot \mathbf{n}$ denotes the component of $\mathbf{v}$ normal to the boundary Γ and $\mathbf{n} \times \mathbf{v} \times \mathbf{n} = \mathbf{v} - (\mathbf{v} \cdot \mathbf{n})\mathbf{n}$ is the projection of $\mathbf{v}$ onto the plane tangent to Γ. In the definition of $\mathbf{V}_0$ we may use $\mathbf{v} \times \mathbf{n} = \mathbf{0}$ due to the relation $\mathbf{v} \times \mathbf{n} = \mathbf{n} \times (\mathbf{n} \times \mathbf{v} \times \mathbf{n})$, i.e., $\mathbf{n} \times \mathbf{v} \times \mathbf{n} = \mathbf{0}$ implies that $\mathbf{n} \times \mathbf{v} = \mathbf{0}$. In $\mathbb{R}^2$, $\mathbf{n} \times \mathbf{v} \times \mathbf{n} = (\mathbf{v} \cdot \tau)\tau$, where τ is the unit tangent vector to Γ.

Suppose that we wish to specify the boundary conditions

$$\mathbf{u} \cdot \mathbf{n} = g_n \text{ on } \Gamma_n \tag{4.6.1}$$

and

$$\mathbf{n} \times \mathbf{u} \times \mathbf{n} = \mathbf{g}_\tau \text{ on } \Gamma_\tau, \tag{4.6.2}$$

i.e., the normal velocity on Γ_n and the tangential velocity on Γ_τ, respectively. For all the weak formulations that we will consider related to any of the possible choices in (4.5), the boundary conditions (4.6) will be *essential boundary conditions*. Thus the trial functions $\mathbf{u}$ will satisfy (4.6), i.e., $\mathbf{u} \in \mathbf{V}_g$, and the test functions $\mathbf{v} \in \mathbf{V}_0$.

Consider the following weak formulations: for $i = 1, 2, 3$ or 4, seek $\mathbf{u} \in \mathbf{V}_g$ and $p \in S$ such that

$$a_i(\mathbf{u}, \mathbf{v}) + b(\mathbf{v}, p) + c(\mathbf{u}, \mathbf{u}, \mathbf{v}) = (\mathbf{f}, \mathbf{v}) + d(\mathbf{v}) \quad \text{for all } \mathbf{v} \in \mathbf{V}_0 \tag{4.7}$$

and

$$b(\mathbf{u}, q) = 0 \quad \text{for all } q \in S. \tag{4.8}$$

Here, $b(\cdot, \cdot)$ and $c(\cdot, \cdot, \cdot)$ remain as in (1.7) and (1.8), respectively, and $\mathbf{f}$ continues to denote the body force appearing in the momentum equation. The linear functional $d(\cdot)$ is given by

$$d(\mathbf{v}) = \int_{\Gamma/\Gamma_n} r\mathbf{v} \cdot \mathbf{n} \, d\Gamma + \int_{\Gamma/\Gamma_\tau} \mathbf{s} \cdot \mathbf{v} \times \mathbf{n} \, d\Gamma, \qquad (4.9)$$

where the functions r and $\mathbf{s}$ are additional data for the problem whose exact physical meaning we will discuss below. In (4.9), for example, Γ/Γ_n denotes the complement of Γ_n in Γ, i.e., $\mathbf{x} \in \Gamma/\Gamma_n$ implies that $\mathbf{x} \in \Gamma$ but $\mathbf{x} \notin \Gamma_n$. Also, since $\mathbf{v}$ is an arbitrary test function, in direction $\mathbf{v} \times \mathbf{n}$ can be taken to be vectors spanning the tangent plane to Γ.

The bilinear forms $a_i(\cdot, \cdot)$, $i = 1, ..., 4$, depend on the choice made in (4.5) and, corresponding to the four possible choices, are given by

$$a_1(\mathbf{u}, \mathbf{v}) = \nu \int_\Omega \operatorname{grad} \mathbf{u} : \operatorname{grad} \mathbf{v} \, d\Omega, \qquad (4.10.1)$$

$$a_2(\mathbf{u}, \mathbf{v}) = \tfrac{1}{2} \int_\Omega \nu(\operatorname{grad} \mathbf{u} + (\operatorname{grad} \mathbf{u})^T) : (\operatorname{grad} \mathbf{v} + (\operatorname{grad} \mathbf{v})^T) \, d\Omega, \quad (4.10.2)$$

$$a_3(\mathbf{u}, \mathbf{v}) = \nu \int_\Omega (\operatorname{curl} \mathbf{u}) \cdot (\operatorname{curl} \mathbf{v}) \, d\Omega, \qquad (4.10.3)$$

and

$$a_4(\mathbf{u}, \mathbf{v}) = \nu \int_\Omega ((\operatorname{curl} \mathbf{u}) \cdot (\operatorname{curl} \mathbf{v}) + (\operatorname{div} \mathbf{u})(\operatorname{div} \mathbf{v})) \, d\Omega. \quad (4.10.4)$$

In the customary manner, should $\mathbf{u}$ and p be sufficiently smooth, one can, through formal integration by parts procedures, ascertain to what differential equation problem the weak formulation (4.7)–(4.8) corresponds. To begin with, we know that the boundary conditions (4.6) are satisfied, since these are being required of the candidate trial functions $\mathbf{u}$. We also find that the differential equations (1.1) and (1.2) are satisfied, where in (1.1) the viscous term is replaced according to (4.5), depending on which choice is made in (4.10). Finally, one finds the *natural boundary conditions* corresponding to the particular weak formulation. We will now discuss these in some detail for each possible choice in (4.10).

Corresponding to the paired choices (4.5.1) and (4.10.1), we have the natural boundary conditions

$$-p + \nu\mathbf{n} \cdot \operatorname{grad} \mathbf{u} \cdot \mathbf{n} = r \text{ on } \Gamma/\Gamma_n \quad \text{and} \quad \nu\mathbf{n} \cdot \operatorname{grad} \mathbf{u} \times \mathbf{n} = \mathbf{s} \text{ on } \Gamma/\Gamma_\tau.$$
$$(4.11.1)$$

These boundary conditions have *no physical meaning*. Thus the choice (4.5.1), or equivalently (4.10.1), can only be used in conjunction with the boundary condition (4.6) specified on *all physical* boundaries. However, for flows with nearly unidirectional outflows, this type of boundary condition, when imposed at these nonphysical outflow boundaries, has been found to yield good computational results; see Glowinski [1984] and Lee *et al.* [1982].

Next, consider the paired choices (4.5.2) and (4.10.2). The natural boundary conditions are then

$$\begin{cases} -p + \nu\mathbf{n} \cdot (\mathrm{grad}\,\mathbf{u} + (\mathrm{grad}\,\mathbf{u})^T) \cdot \mathbf{n} = r \text{ on } \Gamma/\Gamma_n \\ \nu\mathbf{n} \cdot (\mathrm{grad}\,\mathbf{u} + (\mathrm{grad}\,\mathbf{u})^T) \times \mathbf{n} = \mathbf{s} \text{ on } \Gamma/\Gamma_\tau. \end{cases} \quad (4.11.2)$$

Thus r and $\mathbf{s}$ are the normal and tangential stresses, respectively, on the boundary. Then, for the choice (4.10.2), the possible combinations of boundary conditions at a point on the boundary Γ are as follows:

the velocity on $\Gamma_n \cap \Gamma_\tau$,

the stress vector on $\Gamma/(\Gamma_n \cup \Gamma_\tau)$,

the normal velocity and the tangential stress on $\Gamma_n/(\Gamma_n \cap \Gamma_\tau)$, and

the tangential velocity and the normal stress on $\Gamma_\tau/(\Gamma_n \cap \Gamma_\tau)$.

The latter combinations are useful, e.g., for free surface problems or at artificial outflow boundaries.

The third choice, (4.5.3) with (4.10.3), yields the natural boundary conditions

$$-p = r \text{ on } \Gamma/\Gamma_n \quad \text{and} \quad \mathbf{n} \times \boldsymbol{\omega} \times \mathbf{n} = \frac{\mathbf{s}}{\nu} \text{ on } \Gamma/\Gamma_\tau \quad (4.11.3)$$

so that $-r$ and $\mathbf{s}$ are now the pressure p and ν times the tangential vorticity ($\boldsymbol{\omega} = \mathrm{curl}\,\mathbf{u}$), respectively, on the boundary. The possible combinations of boundary conditions are now as follows:

the velocity on $\Gamma_n \cap \Gamma_\tau$,

the pressure and tangential vorticity on $\Gamma/(\Gamma_n \cup \Gamma_\tau)$,

the normal velocity and the tangential vorticity on $\Gamma_n/(\Gamma_n \cap \Gamma_\tau)$, and

the tangential velocity and the pressure on $\Gamma_\tau/(\Gamma_n \cap \Gamma_\tau)$.

The pressure is often used as an outflow condition; the vorticity is useful in exterior problems when matching to an inviscid irrotational flow, since it is well known that the vorticity decays to its far-field value faster than does the velocity.

Unfortunately, although the boundary conditions (4.11.3) associated with (4.10.3) can be useful, in practice it is seldom possible to employ this particular formulation of the viscous term. The reason for this is that the choice (4.10.3) requires that $\mathbf{Z}^h \subset \mathbf{Z}$ in order for the form $a_3(\cdot, \cdot)$ to be coercive on $\mathbf{Z}^h$, i.e., to satisfy (2.9). This condition is, of course, also needed to guarantee the stability of the approximations and, for the other three cases (4.10.1), (4.10.2), and (4.10.4), is trivially satisfied for any choice of conforming discrete velocity space. Therefore the choice (4.10.3) is, in general, not practical.

Fortunately, the boundary conditions (4.11.3) are *approximately* the natural boundary conditions associated with the choice (4.10.4). In fact, for (4.10.4), we have the natural boundary conditions

$$p - v \operatorname{div} \mathbf{u} = r \text{ on } \Gamma/\Gamma_n \quad \text{and} \quad \mathbf{n} \times \omega \times \mathbf{n} = \frac{\mathbf{s}}{v} \text{ on } \Gamma/\Gamma_\tau. \quad (4.11.4)$$

The second of these is identical to the second of (4.11.3). If v is "small," and/or if we assume the incompressibility constraint holds up to portions of the boundary where the normal velocity is *not* specified, then $(p - v \operatorname{div} \mathbf{u})$ is essentially equal to p. Thus we recover, at least approximately, the first boundary condition of (4.11.3).

In summary, when one has velocity and/or stress boundary conditions, one should use (4.11.2) in (4.7), and when one has velocity and/ or pressure and/or vorticity boundary conditions, the choice (4.11.4) is preferable.

The discretization of (4.7)–(4.8) follows the usual procedures once one chooses the finite element spaces for the velocity and the pressure approximations. The natural boundary conditions are automatically accounted for by the inclusion of the linear functions $d(\cdot)$ in (4.7). The essential boundary conditions on the components of the velocity can be enforced in a manner analogous to that described in Section 4.2 for the case where the complete velocity is specified on the whole boundary.

In actuality, there are very few rigorous error estimates available for boundary conditions other than the specification of the velocity. For polygonal or polyhedral domains Ω, the error estimates of Section 2.3

remain valid. However, for domains with *curved boundaries*, using the type of weak formulations discussed here may result in a loss of accuracy. For example, for (4.10.2) with normal velocity and tangential stress boundary conditions, it was shown in Verfurth [1985] that there is a loss of accuracy due to a Babuska type paradox, i.e., the limit of solutions of problems posed on polygonal approximations to $\Omega \subset \mathbb{R}^2$ is not the solution of the problem posed on Ω. In Verfurth [1987a], it is also shown how through the use of additional Lagrange multipliers on the boundary, a different weak formulation yields optimal accuracy.

4.4. The Effects of the Use of Numerical Quadratures

Any implementation of a finite element algorithm requires the evaluation of various integrals. If the viscosity ν is constant, then, except for the integral involving the data $\mathbf{f}$, these integrals may be evaluated exactly, since the integrands are piecewise polynomial functions. However, in practice, all integrals are approximated by quadrature rules. On the one hand, one wishes to use a rule of sufficient accuracy so that the error in the finite element approximation is not adversely affected, at least with respect to the rates of convergence as $h \to 0$. On the other hand, the cost of assembling the discrete system is directly proportional to the number of quadrature points used in the integration rule so that one would like to use as few quadrature points as possible. Here we summarize the work of Crouzeix and Raviart [1973] and Jamet and Raviart [1973] concerning the effect on the error in the finite element approximation resulting from the use of approximate integrations. To keep the exposition manageable, we return to the case discussed in Chapters 1–3, notably of the boundary condition (1.3).

We begin by describing the quadrature rules that are to be used. Assuming, for simplicity, that the domain $\Omega \subset \mathbb{R}^n$ is polyhedral, we let $\mathfrak{I}_h$ denote a triangulation of $\bar{\Omega}$ into n-simplices and we suppose that the discrete velocity space $\mathbf{V}_0^h$ is defined with respect to this triangulation. In addition, we have a discrete pressure space S_0^h defined with respect to a second triangulation $\tilde{\mathfrak{I}}_h$. For example, for the Taylor–Hood element pair we have that $\mathfrak{I}_h = \tilde{\mathfrak{I}}_h$, while for the linear-constant element pairs of Section 3.1, $\mathfrak{I}_h = \tilde{\mathfrak{I}}_{h/2}$.

Let $\hat{\Delta}$ be a reference n-simplex of $\mathbb{R}^n$, e.g., the unit right triangle in $\mathbb{R}^2$. We choose a quadrature rule over the reference set $\hat{\Delta}$ of the form:

$$\mathfrak{I}(\hat{\phi}) = \int_{\hat{\Delta}} \hat{\phi}(\hat{\mathbf{x}}) \, d\hat{\mathbf{x}} \text{ is approximated by } \mathfrak{I}_q(\hat{\phi}) = \sum_{\ell=1}^{L} \hat{\omega}_{\ell} \hat{\phi}(\hat{\mathbf{b}}_{\ell}) \quad (4.12)$$

for some specified quadrature points $\hat{\mathbf{b}}_{\ell} \in \hat{\Delta}$ and weights $\hat{\omega}_{\ell}$, $\ell = 1, ..., L$. For each $\Delta \in \mathfrak{I}_h$, let $\mathbf{F}_{\Delta}: \hat{\mathbf{x}} \to \mathbf{F}_{\Delta}(\hat{\mathbf{x}}) = B_{\Delta}\hat{\mathbf{x}} + \mathbf{a}_{\Delta}$, B_{Δ} an $n \times n$ matrix and $\mathbf{a}_{\Delta}$ an n-vector, be an affine invertible mapping that maps $\hat{\Delta}$ into Δ. We may assume, without loss of generality, that the Jacobian determinant $J_{\Delta} = \det(B_{\Delta})$ of the mapping $\mathbf{F}_{\Delta}$ is positive. Then, corresponding to (4.12) we have the quadrature rule for the set Δ

$$\mathfrak{I}_{\Delta}(\phi) = \int_{\Delta} \phi(\mathbf{x}) \, d\mathbf{x} \text{ is approximated by } \mathfrak{I}_{\Delta q}(\phi) = \sum_{\ell=1}^{L} \omega_{\Delta \ell} \phi(\mathbf{b}_{\Delta \ell}),$$

where

$$\omega_{\Delta \ell} = \hat{\omega}_{\ell} J_{\Delta} \quad \text{and} \quad \mathbf{b}_{\Delta \ell} = \mathbf{F}_{\Delta}(\hat{\mathbf{b}}_{\ell}) \quad \text{for } \ell = 1, ..., L.$$

Then, over the polyhedral domain Ω we have that any integral

$$\mathfrak{I}(\phi) = \int_{\Omega} \phi(\mathbf{x}) \, d\Omega \text{ is approximated by } \mathfrak{I}_q(\phi) = \sum_{\Delta \in \mathfrak{I}_h} \mathfrak{I}_{\Delta q}(\phi). \quad (4.13)$$

Let $a_h(\cdot, \cdot)$, $b_h(\cdot, \cdot)$, $c_h(\cdot, \cdot, \cdot)$, and $(\mathbf{f}, \cdot)_h$ denote the approximations of $a(\cdot, \cdot)$, $b(\cdot, \cdot)$, $c(\cdot, \cdot, \cdot)$, and $(\mathbf{f}, \cdot)$, respectively, resulting from the use of the composite quadrature rule (4.13). For example,

$$c_h(\mathbf{w}, \mathbf{u}, \mathbf{v}) = \mathfrak{I}_q\left(\sum_{i,j=1}^{n} w_j \frac{\partial u_i}{\partial x_j} v_i \right) \quad \text{and} \quad (\mathbf{f}, \mathbf{v})_h = \mathfrak{I}_q\left(\sum_{i=1}^{n} f_i v_i \right).$$

The latter requires the pointwise evaluation of the components of $\mathbf{f}$; thus, we now require that $\mathbf{f} \in \mathbf{C}^0(\bar{\Omega})$. The discrete problem resulting from the use of approximate quadratures is then given by: Seek $\tilde{\mathbf{u}}^h \in \mathbf{V}_0^h$ and $\tilde{p}^h \in S_0^h$ such that

$$a_h(\tilde{\mathbf{u}}^h, \mathbf{v}^h) + c_h(\tilde{\mathbf{u}}^h, \tilde{\mathbf{u}}^h, \mathbf{v}^h) + b_h(\mathbf{v}^h, \tilde{p}^h) = (\mathbf{f}, \mathbf{v}^h)_h \quad \text{for all } \mathbf{v}^h \in \mathbf{V}_0^h \quad (4.14)$$

$$b_h(\tilde{\mathbf{u}}^h, q^h) = 0 \quad \text{for all } q^h \in S_0^h. \quad (4.15)$$

Concerning the discrete spaces $\mathbf{V}_0^h$ and S_0^h, we assume that there exist positive integers k and k' such that

$$\mathbf{P}_k(\Delta) \subset \mathbf{V}_0^h\big|_\Delta \subset \mathbf{P}_{k'}(\Delta) \quad \text{for all } \Delta \in \mathfrak{I}_h, \tag{4.16}$$

$$S_0^h\big|_\Delta \subset P_{k-1}(\Delta) \quad \text{for all } \Delta \in \tilde{\mathfrak{I}}_h, \tag{4.17}$$

$$\text{the triangulation } \mathfrak{I}_h \text{ is regular (see Ciarlet [1978]),} \tag{4.18}$$

and

$$\text{the div-stability condition (2.10) is satisfied.} \tag{4.19}$$

We also assume that the solution $(\mathbf{u}, p)$ satisfies the regularity hypothesis

$$\mathbf{u} \in \mathbf{H}^{k+1}(\Omega) \cap \mathbf{H}_0^1(\Omega) \quad \text{and} \quad p \in H^k(\Omega) \cap L_0^2(\Omega). \tag{4.20}$$

If exact integrations are used to define the discrete system of equations, assumptions (4.16)–(4.20) yield that $|\mathbf{u} - \mathbf{u}^h|_1 + \|p - p^h\|_0 = O(h^k)$ and $\|\mathbf{u} - \mathbf{u}^h\|_0 = O(h^{k+1})$.

We assume that the underlying quadrature rule (4.12) has been chosen so that

$$\hat{\omega}_\ell > 0 \quad \text{for } \ell = 1, ..., L, \tag{4.21}$$

$$\phi \in P_{k'-1} \quad \text{and} \quad \phi(\hat{\mathbf{b}}_\ell) = 0 \quad \text{for } \ell = 1, ..., L \text{ implies that } \phi = 0, \tag{4.22}$$

and

$$\begin{array}{c} \text{the quadrature rule is exact for all polynomials} \\ \text{of degree less than or equal to } k + k' - 2. \end{array} \tag{4.23}$$

The first of these simply states that the quadrature weights in (4.12) are all positive, while (4.22) requires that the set of quadrature points $\{\hat{\mathbf{b}}_\ell\}_{\ell=1}^L$ contains a $P_{k'-1}(\hat{\Delta})$-unisolvent subset, i.e., any polynomial belonging to $P_{k'-1}(\hat{\Delta})$ is uniquely determined by its value on some subset of the quadrature points $\{\hat{\mathbf{b}}_\ell\}_{\ell=1}^L$. Of course, (4.23) is a requirement on the accuracy of the rule employed.

Using the assumptions (4.16)–(4.23), it can then be shown that the solution $(\tilde{\mathbf{u}}^h, \tilde{p}^h)$ of the discrete system (4.14)–(4.15) satisfies the estimate

$$|\mathbf{u} - \tilde{\mathbf{u}}^h|_1 + \|p - \tilde{p}^h\|_0 = O(h^k). \tag{4.24}$$

Thus, the same rate of convergence is obtained as for the solution $(\mathbf{u}^h, p^h)$ of the discrete system (2.1)–(2.2) that is obtained through the use of exact integrations. If we make the further assumptions that

$k' \leq k + 1$ and that the quadrature rule (4.12) is exact for all polynomials of degree less than or equal to $\max(k' + k - 2, k)$, we then have that

$$\|u - \tilde{u}^h\|_0 = O(h^{k+1}). \tag{4.25}$$

Again the same rate of convergence is obtained as was the case for u^h.

In order to illustrate these results, let us first consider the Taylor–Hood element pair for which $k = k' = 2$ so that $k + k' - 2 = 2$. We choose the midside quadrature rule for which the quadrature points for any triangle are the three edge midpoints and the quadrature weights are one-third the area of the triangle. This rule is exact for all polynomials of degree less than or equal to two so that if (4.18) and (4.20) hold, one then obtains the estimates (4.24) and (4.25) with $k = 2$. Thus the rates of convergence of the Taylor–Hood element pair are preserved if one merely uses the midside quadrature rule in defining the discrete system of equations. Note that with this rule, the viscous term and the pressure term, both of which involve at most quadratic polynomials, are integrated exactly, i.e., $a_h(u^h, v^h) = a(u^h, v^h)$ and $b_h(v^h, p^h) = b(v^h, p^h)$ for all $u^h, v^h \in V_0^h$ and $p^h \in S_0^h$. However, the convection term, which involves quintic polynomials, is not integrated exactly by the midside quadrature rule, i.e., $c_h(u^h, u^h, v^h) \neq c(u^h, u^h, v^h)$. The midside quadrature rule may be used for the modified Taylor–Hood element of Thatcher and Silvester [1987] (see Section 3.6) without affecting the rate of convergence of the approximations.

Next, let us consider piecewise linear velocity elements in conjunction with piecewise constant pressures. Now $k = k' = 1$ so that $k + k' - 2 = 0$. Therefore, we may use the one-point centroid rule for which the single quadrature point in each triangle is the centroid of the triangle and the quadrature weight is the area of the triangle. This one point rule is exact for linear functions, and again the viscous and pressure terms are integrated exactly. On the other hand, the convection term involves quadratic polynomials and is not integrated exactly. Still, using this one-point rule, we preserve the $O(h)$ error estimates for the pressure in the $L^2(\Omega)$-norm and the velocity in the $H^1(\Omega)$-norm. Also, $\max(k + k' - 2, k) = 1$ so that the hypotheses necessary for obtaining (4.25) are also satisfied; hence (4.25) provides an $O(h^2)$ estimate for the velocity error in the $L^2(\Omega)$-norm. Note that the theory does not apply to the linear velocity–linear pressure element pair since in this case $S_0^h|_\Delta \not\subset P_{k-1}(\Delta)$. However, since the one-point centroid rule still yields that $b_h(\mathbf{v}^h, p^h) = b(\mathbf{v}^h, p^h)$, it is probable that this rule is sufficiently accurate to obtain the error estimates (4.25)–(4.26) with $k = 1$.

We may also consider some of the bubble elements of Section 3.3. For example, consider the velocity space where a continuous piecewise quadratic space is augmented, in each triangle, with the cubic bubble function that vanishes along the three edges of the triangle. The corresponding pressure space consists of all discontinuous piecewise linear functions. Here we have $k = 2$ and, since cubic functions are present, $k' = 3$, so that $k + k' - 2 = 3$. Now the midside quadrature rule does not suffice to preserve the error estimates obtainable through the use of exact integrations. Indeed, we now need a quadrature rule that integrates cubic polynomials exactly. Note that the pressure term, i.e., $b(\mathbf{v}^h, p^h)$, that involves at most cubic polynomials is integrated exactly, but neither the viscous term or the convection term, which may involve up to quartic and eight degree polynomials, respectively, are integrated exactly.

One rule of thumb may be gleaned from the above results and examples. In order to preserve the accuracy of the finite element approximations, it suffices to use a quadrature rule that integrates the viscous term exactly. In many cases, e.g., the Taylor–Hood element pair, this rule of thumb gives the minimally accurate quadrature rule that preserves the accuracy of the finite element approximation; in other cases, i.e., the above bubble element example, this rule of thumb indicates an unnecessarily accurate rule, but in any case, yields a rule that preserves accuracy.

5

Penalty Methods

Penalty methods provide a means for uncoupling the determination of the velocity and pressure. The mechanism that allows for this uncoupling is the relaxation of the incompressibility constraint. This method has gained substantial popularity in engineering circles for at least three reasons. First, the uncoupling of the velocity and pressure leads to a reduction in the number of degrees of freedom one has to solve for; also, the incompressibility condition can be removed from the solution process for the velocity and therefore, as far as velocity calculations are concerned, the div-stability condition may be ignored; and, finally, because it works well in practice, i.e., it produces good answers in an efficient manner. The fact that the div-stability condition is evidently circumvented allows one to view penalty methods as regularization methods.

Penalty methods are, of course, an old adjunct to the method of Lagrange multipliers. In the context of the Stokes and Navier–Stokes equations, the method goes as far back as Temam [1968]. There is, by now, a vast literature on penalty methods in incompressible flow problems, both in and out of the finite element framework. For example, see Bercovier [1978], Falk [1975], Girault and Raviart [1979 and 1986], Pelissier [1975], and Temam [1979] for mathematical discussions, and Cuvelier, Segal, and van Steenhoven [1986] and Hughes, Liu, and Brooks [1979] for thorough discussions of implementation issues.

5.1. The Penalty Method for the Continuous Problem and Its Discretization

Let's begin by explaining the use of the terminology *penalty* in the name of the method. It is well known (see, e.g., Girault and Raviart [1986]) that the Stokes problem

$$-\nu\,\Delta\mathbf{u} + \operatorname{grad} p = \mathbf{f} \text{ in } \Omega, \tag{5.1}$$

$$\operatorname{div}\mathbf{u} = 0 \text{ in } \Omega, \tag{5.2}$$

and

$$\mathbf{u} = \mathbf{0} \text{ on } \Gamma \tag{5.3}$$

can be equivalently formulated as a minimization problem. In fact, consider the functional

$$\mathfrak{J}(\mathbf{v}) = \nu \int_\Omega \operatorname{grad}\mathbf{v} : \operatorname{grad}\mathbf{v}\,d\Omega - 2\int_\Omega \mathbf{f}\cdot\mathbf{v}\,d\Omega = a(\mathbf{v},\mathbf{v}) - 2(\mathbf{f},\mathbf{v}),$$

where $a(\cdot,\cdot)$ is defined in (1.6) and also the minimization problem

$$\{\min \mathfrak{J}(\mathbf{v}) : \operatorname{div}\mathbf{v} = 0 \text{ in } \Omega \text{ and } \mathbf{v} = \mathbf{0} \text{ on } \Gamma\}, \tag{5.4}$$

where at this point we are proceeding in a formal manner. With the introduction of the Lagrange multiplier p to enforce the incompressibility constraint, it is easy to see that the Stokes equations (5.1)–(5.3) are the Euler–Lagrange equations for the minimization problem,

$$\{\min \tilde{\mathfrak{J}}(\mathbf{v}, p) : \mathbf{v} = \mathbf{0} \text{ on } \Gamma\},$$

where

$$\tilde{\mathfrak{J}}(\mathbf{v}, p) = a(\mathbf{v},\mathbf{v}) - 2(\mathbf{f},\mathbf{v}) - 2(\operatorname{div}\mathbf{v}, p) = a(\mathbf{v},\mathbf{v}) - 2(\mathbf{f},\mathbf{v}) + 2b(\mathbf{v}, p)$$

and where $b(\cdot,\cdot)$ is defined in (1.7). It is well known that one may also enforce a constraint in a minimization problem through penalization. Thus, let $\varepsilon > 0$ be a given number, and let

$$\mathfrak{J}_\varepsilon(\mathbf{v}) = a(\mathbf{v},\mathbf{v}) - 2(\mathbf{f},\mathbf{v}) - \frac{1}{\varepsilon}(\operatorname{div}\mathbf{v}, \operatorname{div}\mathbf{v}).$$

Then the minimization problem

$$\min\{\mathcal{J}_\varepsilon(\mathbf{v}) : \mathbf{v} = \mathbf{0} \text{ on } \Omega\} \tag{5.5}$$

is a penalized version of the minimization problem (5.4). Now the incompressibility condition (5.2) can be enforced as well as one likes (but not exactly) by choosing ε sufficiently small.

If we denote by $\mathbf{u}_\varepsilon$ the solution of (5.5), we have that the Euler–Lagrange equations for (5.5) are given by

$$-\nu \, \Delta \mathbf{u}_\varepsilon - \frac{1}{\varepsilon} \text{grad}(\text{div } \mathbf{u}_\varepsilon) = \mathbf{f} \text{ in } \Omega \tag{5.6}$$

and

$$\mathbf{u}_\varepsilon = 0 \text{ on } \Gamma. \tag{5.7}$$

If we introduce $p_\varepsilon = -(\text{div } \mathbf{u}_\varepsilon)/\varepsilon$, then (5.6)–(5.7) can be written in the form

$$-\nu \, \Delta \mathbf{u}_\varepsilon + \text{grad } p_\varepsilon = \mathbf{f} \text{ in } \Omega, \tag{5.8}$$

$$\text{div } \mathbf{u}_\varepsilon = -\varepsilon p_\varepsilon \text{ in } \Omega, \tag{5.9}$$

and

$$\mathbf{u}_\varepsilon = 0 \text{ on } \Gamma. \tag{5.10}$$

Comparing with the Stokes equations (5.1)–(5.3), we see that the effect of penalization is simply to relax the incompressibility condition (5.2). Also, as $\varepsilon \to 0$ in (5.8)–(5.10), we recover the Stokes equations (5.1)–(5.3).

The above discussion explains the use of the terminology *penalty*. However, it is more common to view things in reverse order. Thus, one starts out by viewing (5.8)–(5.10) as a perturbation of the Stokes equations (5.1)–(5.3). Then, one can easily eliminate the pressure from the system (5.8)–(5.10) to arrive at the system (5.6)–(5.7). This second viewpoint immediately generalizes to the nonlinear Navier–Stokes equations. Given a parameter ε, we replace (1.1)–(1.3) by the perturbed version

$$-\nu \, \Delta \mathbf{u}_\varepsilon + \mathbf{u}_\varepsilon \cdot \text{grad } \mathbf{u}_\varepsilon + \text{grad } p_\varepsilon = \mathbf{f} \text{ in } \Omega, \tag{5.11}$$

$$\text{div } \mathbf{u}_\varepsilon = -\varepsilon p_\varepsilon \text{ in } \Omega, \tag{5.12}$$

and

$$\mathbf{u}_\varepsilon = 0 \text{ on } \Gamma. \tag{5.13}$$

Again, we may easily eliminate the pressure to yield

$$-\nu \Delta \mathbf{u}_\varepsilon + \mathbf{u}_\varepsilon \cdot \text{grad } \mathbf{u}_\varepsilon - \frac{1}{\varepsilon} \text{grad}(\text{div } \mathbf{u}_\varepsilon) = \mathbf{f} \text{ in } \Omega \qquad (5.14)$$

and

$$\mathbf{u}_\varepsilon = 0 \text{ on } \Gamma. \qquad (5.15)$$

If ε is small, it is clear that the penalized equations (5.8)–(5.10) and
(5.11)–(5.13) are "close" to the unpenalized equations (5.1)–(5.3) and
(1.1)–(1.3), respectively. It would be nice if the solutions of the corresponding equations were close as well. In fact, it can be shown, for
solutions of the Stokes equations and for nonsingular branches of
solutions in the Navier–Stokes equations, that

$$|\mathbf{u} - \mathbf{u}_\varepsilon|_1 + \|p - p_\varepsilon\|_0 \le C\varepsilon \qquad (5.16)$$

where the constant C is independent of ε. Thus, as $\varepsilon \to 0$, the solutions
of the penalized problems converge to those of the unpenalized
problems.

At this point we may discretize (5.6)–(5.7) or (5.14)–(5.15) by standard
techniques used for second-order elliptic partial differential equations;
see, e.g., Babuska and Aziz [1972], Ciarlet [1978], or Strang and Fix
[1973]. For example, consider the following weak formulation. Given
$\mathbf{f} \in \mathbf{H}^{-1}(\Omega)$, find $\mathbf{u}_\varepsilon \in \mathbf{H}_0^1(\Omega)$ such that

$$a(\mathbf{u}_\varepsilon, \mathbf{v}) + \frac{1}{\varepsilon}(\text{div } \mathbf{u}_\varepsilon, \text{div } \mathbf{v}) + c(\mathbf{u}_\varepsilon, \mathbf{u}_\varepsilon, \mathbf{v}) = (\mathbf{f}, \mathbf{v}) \quad \text{for all } \mathbf{v} \in \mathbf{H}_0^1(\Omega),$$
$$(5.17)$$

where $c(\cdot, \cdot, \cdot)$ is defined in (1.8). One easily sees that solutions of
(5.17) are weak solutions of (5.14)–(5.15). In order to discretize (5.17)
one merely has to choose a finite element subspace $\mathbf{V}_0^h \subset \mathbf{H}_0^1(\Omega)$ and
find a $\mathbf{u}_\varepsilon^h \in \mathbf{V}_0^h$ such that

$$a(\mathbf{u}_\varepsilon^h, \mathbf{v}^h) + \frac{1}{\varepsilon}(\text{div } \mathbf{u}_\varepsilon^h, \text{div } \mathbf{v}^h) + c(\mathbf{u}_\varepsilon^h, \mathbf{u}_\varepsilon^h, \mathbf{v}^h) = (\mathbf{f}, \mathbf{v}^h) \quad \text{for all } \mathbf{v}^h \in \mathbf{V}_0^h.$$
$$(5.18)$$

Then, it is not hard to show that on branches of nonsingular solutions

$$|\mathbf{u}_\varepsilon - \mathbf{u}_\varepsilon^h|_1 \le Ch^s, \qquad (5.19)$$

where the exponent s depends in the usual way on the regularity of the
solution and the approximating properties of $\mathbf{V}_0^h$. The estimates (5.16)

and (5.19) may be combined to yield

$$|\mathbf{u} - \mathbf{u}_\varepsilon^h|_1 \le C(h^s + \varepsilon). \tag{5.20}$$

Thus, by choosing ε sufficiently small, one can achieve the same accuracy for the velocity as one does in the unpenalized method.

One may now ask what happend to the div-stability condition? Up to this point it has not entered the discussion, and in obtaining the error estimate (5.20) it plays no role, i.e., there has been no constraint placed on the choice for the velocity finite element space $\mathbf{V}_0^h$. But now we turn to the computed pressure, which naturally we define to be

$$p_\varepsilon^h|_\Delta = -\frac{1}{\varepsilon}\,\mathrm{div}(\mathbf{u}_\varepsilon^h|_\Delta), \tag{5.21}$$

where Δ denotes any element in the subdivision of Ω into finite elements. Therefore, implicitly we are defining a pressure finite element space S_0^h that consists of functions that, in each element, are the divergence of some function belonging to $\mathbf{V}_0^h$. [Note that indeed $S_0^h \subset L_0^2(\Omega)$ because of (5.21), the divergence theorem, and the fact that $\mathbf{u}_\varepsilon^h \in \mathbf{H}_0^1(\Omega)$.] It turns out that if the pair of spaces $\mathbf{V}_0^h$ and S_0^h does not satisfy the div-stability condition, then the discrete pressure computed by (5.21) will not be very good, e.g., it will exhibit spurious oscillations; see Cuvelier, Segal, and van Steenhoven [1986]. However, in practice, even such pressures may be smoothed so that meaningful approximations can be obtained; see Hughes, Liu, and Brooks [1979].

5.2. The Discrete Penalty Method

Our approach so far has been to uncouple the pressure from the velocity before discretizing. One may also reverse the order of these steps. There is some advantage in doing so in that one then has control over the choice for the pressure finite element space, i.e., it will not be dictated by the relation (5.21).

Thus we start with the discrete Navier–Stokes equations [see (2.1)–(2.2)]

$$a(\mathbf{u}^h, \mathbf{v}^h) + b(\mathbf{v}^h, p^h) + c(\mathbf{u}^h, \mathbf{u}^h, \mathbf{v}^h) = (\mathbf{f}, \mathbf{v}^h) \quad \text{for all } \mathbf{v}^h \in \mathbf{V}_0^h \tag{5.22}$$

and

$$b(\mathbf{u}^h, q^h) = 0 \quad \text{for all } q^h \in S_0^h, \tag{5.23}$$

where $\mathbf{u}^h \in \mathbf{V}_0^h$ and $p^h \in S_0^h$. We assume that the pair of finite element spaces $\mathbf{V}_0^h$ and S_0^h satisfies the div-stability condition so that the error estimate of the type

$$|\mathbf{u} - \mathbf{u}^h|_1 + \|p - p^h\|_0 \le Ch^s \qquad (5.24)$$

holds for some constant s; see Chapter 3.

The penalized or perturbed version of (5.22)–(5.23) bears the same relation to these equations as do (5.11)–(5.13) to the Navier–Stokes equations. Thus, we seek $\mathbf{u}_\varepsilon^h \in \mathbf{V}_0^h$ and $p_\varepsilon^h \in S_0^h$ such that

$$a(\mathbf{u}_\varepsilon^h, \mathbf{v}^h) + b(\mathbf{v}^h, p_\varepsilon^h) + c(\mathbf{u}_\varepsilon^h, \mathbf{u}_\varepsilon^h, \mathbf{v}^h) = (\mathbf{f}, \mathbf{v}^h) \qquad \text{for all } \mathbf{v}^h \in \mathbf{V}_0^h \quad (5.25)$$

and

$$b(\mathbf{u}_\varepsilon^h, q^h) - \varepsilon(p_\varepsilon^h, q^h) = 0 \qquad \text{for all } q^h \in S_0^h. \qquad (5.26)$$

Note that $\mathbf{u}_\varepsilon^h$ and p_ε^h that satisfy (5.25)–(5.26) are not the same as those that satisfy (5.18) and (5.21).

By the definition of $\mathcal{P}_h$, the orthogonal projection operator of $L^2(\Omega)$ onto S_0^h, we have from (5.26) that $p_\varepsilon^h = -[\mathcal{P}_h(\operatorname{div} \mathbf{u}_\varepsilon^h)]/\varepsilon$. This is a discrete analogue of (5.12) and allows us to eliminate p_ε^h from (5.25). In fact, we have that

$$a(\mathbf{u}_\varepsilon^h, \mathbf{v}^h) + \frac{1}{\varepsilon}(\operatorname{div} \mathbf{v}^h, \mathcal{P}_h(\operatorname{div} \mathbf{u}_\varepsilon^h)) + c(\mathbf{u}_\varepsilon^h, \mathbf{u}_\varepsilon^h, \mathbf{v}^h) = (\mathbf{f}, \mathbf{v}^h)$$

$$\text{for all } \mathbf{v}^h \in \mathbf{V}_0^h. \qquad (5.27)$$

Note the difference between (5.18) and (5.27).

To help clarify the notions discussed above, let us restate them in matrix notation. To this end we let $\mathbf{U}$ and $\mathbf{P}$, respectively, denote the vectors of velocity and pressure degrees of freedom, e.g., nodal values. Then one can easily see that (5.25)–(5.26) are equivalent to the nonlinear problem

$$A\mathbf{U} + N(\mathbf{U})\mathbf{U} + B^T\mathbf{P} = \mathbf{F} \qquad (5.28)$$

$$B\mathbf{U} - \varepsilon M\mathbf{P} = 0, \qquad (5.29)$$

where $-B$ is a discrete divergence matrix, B^T is a discrete gradient matrix, M is the mass or Gram matrix for the basis used for S_0^h, $-A$ is a vector Laplacian matrix (multiplied by ν), $N(\mathbf{U})\mathbf{U}$ represents the nonlinear convection term, and $\mathbf{F}$ is a data vector. The matrices A and M are symmetric and positive definite; thus, one can easily eliminate

P from (5.28)–(5.29) to yield

$$\left(A + \frac{1}{\varepsilon} B^T M^{-1} B\right)\mathbf{U} + N(\mathbf{U})\mathbf{U} = \mathbf{F}. \tag{5.30}$$

On the other hand, using the same velocity finite element space, the matrix form of (5.18) looks like

$$\left(A + \frac{1}{\varepsilon} S\right)\mathbf{U} + N(\mathbf{U})\mathbf{U} = \mathbf{F}, \tag{5.31}$$

where S is a symmetric positive semidefinite matrix equivalent to the penalty term in (5.18). Certainly $S \neq B^T M^{-1} B$ so that the differences in the two approaches to the penalty method are now evident.

The necessity of inverting M in (5.30) seems to put the approach wherein one discretizes first and then penalizes at a disadvantage over the reverse approach leading to (5.31); the latter does not involve any matrix inversions. However, for discontinuous pressure spaces, the inversion of M can be done locally, i.e., one element at a time, so that having to compute M^{-1} is no real disadvantage. For this reason, the penalty method is especially effective when used in conjunction with discontinous pressure spaces. It should also be pointed out that there are those who wait to penalize until they are at the matrix stage, i.e., they solve the discrete Navier–Stokes equations, which in matrix form are given by (5.28)–(5.29) with $\varepsilon = 0$, by adding the perturbation $\varepsilon D\mathbf{P}$, where D is a diagonal matrix, to the discrete continuity equation; this is, of course, equivalent to approximating the mass matrix M by the diagonal matrix D, a process known as *lumping*.

As was the case for the continuous problem, one can estimate the difference between the solution of the penalized discrete problem (5.25)–(5.26), or equivalently, (5.27), and the exact solution of the discrete Navier–Stokes equations (5.22)–(5.23). The result is

$$|\mathbf{u}^h - \mathbf{u}_\varepsilon^h|_1 + \|p^h - p_\varepsilon^h\|_0 \le C\varepsilon.$$

Combining with (5.24) yields an estimate of the form

$$|\mathbf{u} - \mathbf{u}_\varepsilon^h|_1 + \|p - p_\varepsilon^h\|_0 \le C(\varepsilon + h^s).$$

Thus, again, by choosing ε sufficiently small, the second approach to penalty methods does not harm the accuracy of the approximation.

It is worth spending some time exploring the role the div-stability condition plays within penalty methods leading to (5.27) or its matrix analogue, (5.30). In this case, we have assumed that the div-stability condition holds in order to derive the error estimate (5.24); however, it is not obvious what div-stability has to do with equation (5.27). Recall that the pressure is given by $p_\varepsilon^h = -[\mathcal{P}_h(\text{div } \mathbf{u}_\varepsilon^h)]/\varepsilon$, or in terms of matrices, $\mathbf{P} = \varepsilon^{-1} M^{-1} B \mathbf{U}$. The div-stability condition implies that the rectangular matrix B is of full rank equal to the number of its rows. Now examine $B^T \mathbf{P}$, which is the discrete gradient operator acting on the discrete pressure. Now, $B^T \mathbf{P} = \varepsilon^{-1} B^T M^{-1} B \mathbf{U}$ and the matrix $B^T M^{-1} B$ is positive definite whenever B is of full rank. Thus when the div-stability condition is satisfied B is of full rank and the null space of the discrete gradient operator contains only the zero vector. [Recall that we have removed the possibility of nonzero constant pressures by constraining the pressure space to be a subspace of $L_0^2(\Omega)$.] The role of the div-stability condition is to rule out spurious pressure modes, e.g., the checkerboard mode discussed in Section 2.5. It probably plays no role in the stability or accuracy of velocity approximations obtained from (5.27).

A question that arises in practice is choosing the penalty parameter ε. Of course, one would want to choose this as small as possible, but one must watch out for problems such as the effects of round-off errors if one chooses ε too small. See Hughes, Liu, and Brooks [1979] for a thorough discussion of this issue.

It is easily seen that penalty methods may be viewed as the first step of an iterated penalty, or, in the Stokes equations case, an augmented Lagrangian method. Consider the following iteration. Given $p^{(0)}$, determine the sequence $\{\mathbf{u}^{(k)}, p^{(k)}\}$, $k \geq 1$, from

$$-\nu \Delta \mathbf{u}^{(k)} + \mathbf{u}^{(k)} \cdot \text{grad } \mathbf{u}^{(k)} + \text{grad } p^{(k)} = \mathbf{f} \text{ in } \Omega, \tag{5.32}$$

$$\text{div } \mathbf{u}^{(k)} + \varepsilon p^{(k)} = +\varepsilon p^{(k-1)} \text{ in } \Omega, \tag{5.33}$$

and

$$\mathbf{u}^{(k)} = \mathbf{0} \text{ on } \Gamma. \tag{5.34}$$

Clearly, if $p^{(0)} = 0$, these equations, with $k = 1$, are equivalent to (5.11)–(5.13) and therefore, in this case, $\mathbf{u}^{(1)} = \mathbf{u}_\varepsilon$. However, the iteration (5.32)–(5.34) produces ever improving approximations (compared to the penalty method approximation $\mathbf{u}_\varepsilon$); in fact, one can show that

$$|\mathbf{u} - \mathbf{u}^{(k)}|_1 + \|p - p^{(k)}\|_0 = O(\varepsilon^k);$$

one should compare this result with (5.16). See Glowinski [1984] and

Segal [1979] for details in the Stokes flow setting and Fortin and Fortin [1985] for the Navier–Stokes setting.

All of our comments hold in the simpler setting of the Stokes equations; this is noteworthy because one common use of the penalty method is for solving discrete Stokes problems that arise within some solution algorithm for the nonlinear discrete Navier–Stokes equations. Also, the penalty methods discussed here can be extended to problems with other boundary conditions, including inhomogeneous ones. For time-dependent problems one may also define penalty methods; see, e.g., Cuvelier, Segal, and van Steenhoven [1986] or Hughes, Liu, and Brooks [1979]. A related method is the *artificial compressibility method* wherein one replaces the continuity equation div $\mathbf{u} = 0$ by the equation

$$-\varepsilon \frac{\partial p_\varepsilon}{\partial t} = \operatorname{div} \mathbf{u}_\varepsilon.$$

See, e.g., Chorin [1967], Peyret, and Taylor [1983]; Temam [1979]; and Yanenko [1971).

II

SOLUTION OF THE DISCRETE EQUATIONS

In Chapters 1-5 we examined finite element discretizations of the Navier–Stokes equations; we now consider methods for the solution of the resulting discrete equations. From both the algorithmic and analysis points of view, the situation here is not so good as it is for the discretization step. For two-dimensional flows at low values of the Reynolds number, there exist numerous solution algorithms that work very well; however, for flows in three dimensions and/or for flows at high values of the Reynolds number, there still needs to be efficient algorithms developed and analyzed for the solution of discrete equations. Indeed, this is the most important outstanding problem concerning the approximate solution of the Navier–Stokes equations and should continue to provide fertile research ground for some time to come, especially in light of the emergence of novel computer architectures.

6

Newton's Method and Other Iterative Methods

6.1. Newton's Method

The discrete system resulting from a finite element discretization of the Navier-Stokes equations constitutes a nonlinear system of algebraic equations. These are given explicitly by (2.3)–(2.4); however, in order to simplify our exposition we will, for the most part, discuss iterative methods using the more symbolic representation (2.1)–(2.2). Thus we seek $\mathbf{u}^h \in \mathbf{V}_0^h$ and $p^h \in S_0^h$ such that

$$a(\mathbf{u}^h, \mathbf{v}^h) + c(\mathbf{u}^h, \mathbf{u}^h, \mathbf{v}^h) + b(\mathbf{v}^h, p^h) = (\mathbf{f}, \mathbf{v}^h) \quad \text{for all } \mathbf{v}^h \in \mathbf{V}_0^h \quad (6.1)$$

and

$$b(\mathbf{u}^h, q^h) = 0 \quad \text{for all } q^h \in S_0^h. \quad (6.2)$$

We assume that the finite element spaces $\mathbf{V}_0^h$ and S_0^h satisfy the div-stability condition (2.10) as well as the conditions (2.5)–(2.7) and (2.9) so that the discrete solution $(\mathbf{u}^h, p^h)$ is optimally accurate in the sense of (2.16) and (2.17).

Newton's method for the approximate solution of (6.1)–(6.2) is described as follows. Given an initial guess $\mathbf{u}^{(0)} \in \mathbf{V}_0^h$, one generates the sequence $\{\mathbf{u}^{(m)}, p^{(m)}\}$ for $m = 1, 2, 3, \ldots$ by solving the sequence

81

of *linear* problems

$$a(\mathbf{u}^{(m)}, \mathbf{v}^h) + c(\mathbf{u}^{(m)}, \mathbf{u}^{(m-1)}, \mathbf{v}^h) + c(\mathbf{u}^{(m-1)}, \mathbf{u}^{(m)}, \mathbf{v}^h) + b(\mathbf{v}^h, p^{(m)})$$
$$= (\mathbf{f}, \mathbf{v}^h) + c(\mathbf{u}^{(m-1)}, \mathbf{u}^{(m-1)}, \mathbf{v}^h) \quad \text{for all } \mathbf{v}^h \in \mathbf{V}_0^h \tag{6.3}$$

and

$$b(\mathbf{u}^{(m)}, q^h) = 0 \quad \text{for all } q^h \in S_0^h. \tag{6.4}$$

Note that the initial guess $\mathbf{u}^{(0)}$ is required to satisfy the boundary condition $\mathbf{u}^{(0)} = \mathbf{0}$ on Γ but is not required to satisfy the incompressibility constraint, even in its weak form (6.2). Also note that no initial guess for the pressure need be supplied.

Given the basis sets $\{q_j(\mathbf{x})\}, j = 1, \dots, J$, and $\{\mathbf{v}_k(\mathbf{x})\}, k = 1, \dots, K$, for S_0^h and $\mathbf{V}_0^h$, respectively, one may write

$$p^{(m)} = \sum_{j=1}^{J} \alpha_j^{(m)} q_j(\mathbf{x}) \quad \text{and} \quad \mathbf{u}^{(m)} = \sum_{k=1}^{K} \beta_k^{(m)} \mathbf{v}_k(\mathbf{x}) \tag{6.5}$$

for some constants $\alpha_j^{(m)}, j = 1, \dots, J$, and $\beta_k^{(m)}, k = 1, \dots, K$. Then, the linear system (6.3)–(6.4) that determines $p^{(m)}$ and $\mathbf{u}^{(m)}$ is explicitly given by

$$\sum_{k=1}^{K} A_{\ell k}^{(m)} \beta_k^{(m)} + \sum_{j=1}^{J} B_{j\ell} \alpha_j^{(m)} = F_\ell^{(m)} \quad \text{for } \ell = 1, \dots, K \tag{6.6}$$

and

$$\sum_{k=1}^{K} B_{jk} \beta_k^{(m)} = 0 \quad \text{for } j = 1, \dots, J, \tag{6.7}$$

where

$$A_{\ell k}^{(m)} = a(\mathbf{v}_k, \mathbf{v}_\ell) + c(\mathbf{v}_k, \mathbf{u}^{(m-1)}, \mathbf{v}_\ell) + c(\mathbf{u}^{(m-1)}, \mathbf{v}_k, \mathbf{v}_\ell)$$
$$\text{for } k, \ell = 1, \dots, K, \tag{6.8}$$

$$B_{jk} = b(\mathbf{v}_k, q_j) \quad \text{for } j = 1, \dots, J \text{ and } k = 1 \dots, K, \tag{6.9}$$

and

$$F_\ell^{(m)} = (\mathbf{f}, \mathbf{v}_\ell) + c(\mathbf{u}^{(m-1)}, \mathbf{u}^{(m-1)}, \mathbf{v}_\ell) \quad \text{for } \ell = 1, \dots, K. \tag{6.10}$$

The coefficient matrix of the linear system (6.6)–(6.7) depends on $\mathbf{u}^{(m-1)}$; thus each Newton iterate is determined as the solution of a *different* linear system.

The first question that needs to be addressed is whether or not the Newton iterates $\{\mathbf{u}^{(m)}, p^{(m)}\}$ are well defined, i.e., is the Jacobian matrix of the linear system (6.3)–(6.4), or equivalently, of (6.6)–(6.7), nonsingular? One may then ask what conditions are sufficient to guarantee the convergence of the iterates to a solution of (6.1)–(6.2)? Finally, what is the rate of convergence of the iterates? Unlike many other methods for the solution of the discrete system (6.1)–(6.2), Newton's method is guaranteed (see, e.g., Girault and Raviart [1986]) to be locally and quadratically convergent whenever the initial guess $\mathbf{u}^{(0)}$ for the velocity is "sufficiently close" to a branch of nonsingular solutions. We now elaborate on these points.

The most extensive available results (Girault and Raviart [1979 and 1986], Gunzburger and Peterson [1983], and Karakashian [1982]) apply to the case where one can guarantee that the Navier–Stokes equations have a unique solution, i.e., (2.13) holds. Let

$$d = \frac{\nu}{2\kappa_c} \left(1 - \frac{\kappa_c \|\mathbf{f}\|_{-1}}{\nu^2}\right) \tag{6.11}$$

and suppose that the initial guess satisfies

$$\|\mathbf{u}^{(0)} - \mathbf{u}^h\|_1 \le d \quad \text{and} \quad \mathbf{u}^{(0)} = 0 \text{ on } \Gamma. \tag{6.12}$$

Then the sequence of Newton iterates $\{\mathbf{u}^{(m)}, p^{(m)}\}$, $m = 1, 2, \ldots$, are uniquely defined by (6.3)–(6.4), or equivalently, by (6.5)–(6.10) and $|\mathbf{u}^{(m)} - \mathbf{u}^h|_1 \le d$ for all m. In fact, for $m \ge 1$

$$|\mathbf{u}^{(m)} - \mathbf{u}^h|_1 \le \frac{|\mathbf{u}^{(0)} - \mathbf{u}^h|_1^{2^m}}{d^{2^m - 1}}. \tag{6.13}$$

Furthermore, if $|\mathbf{u}^{(0)} - \mathbf{u}^h|_1 = d\varepsilon$ with $0 < \varepsilon < 1$, then for $m \ge 1$

$$|\mathbf{u}^{(m)} - \mathbf{u}^h|_1 \le d\varepsilon^{2^m} \tag{6.14}$$

and, for some constant $C > 0$,

$$\|p^{(m)} - p^h\|_0 \le C\varepsilon^{2^m}. \tag{6.15}$$

The results (6.13)–(6.15) indicate that whenever (6.11) is satisfied and the initial guess satisfies (6.12), then the subsequent Newton iterates are well defined and converge quadratically to the unique solution of (6.1)–(6.2).

In the more general situation of the solution of (6.1)–(6.2) belonging to a nonsingular branch, one may still show (Girault and Raviart [1986]) that, should the initial guess $\mathbf{u}^{(0)}$ be sufficiently close to $\mathbf{u}^h$, then the Newton iterates converge quadratically. For example, note the similarity between the left-hand sides of (2.14) and (6.1)–(6.2), so that on a nonsingular branch the Jacobian coefficient matrix of (6.3)–(6.4) is invertible, i.e., the Newton iterates are well defined.

The main advantage that Newton's method enjoys over other iterative methods for the solution of (6.1)–(6.2) is its locally quadratic convergence. On the other hand, a new linear system of algebraic equations must be solved at each step of the Newton iteration. A further disadvantage of Newton's method is that convergence occurs only for sufficiently close initial guesses. Fortunately, the situation is not so bad as predicted by (6.11)–(6.12), i.e., d is a poor estimate for the radius of the attraction ball of Newton's method. However, it is found in practice that this radius does decrease with increasing Reynolds number, or with decreasing ν. Thus, as the Reynolds number $1/\nu$ increases, better initial guesses are needed in order for the Newton iteration to converge. In the following discussion, other iterative methods are discussed that avert some of these disadvantages of Newton's method; of course, the price paid is that the quadratic convergence property is lost. We will also discuss the generation of good initial guesses for Newton's and other's methods.

6.2. A Fixed Jacobian Method

One obvious way to avoid solving a different linear system at each step of an iterative process is to keep the coefficient matrix fixed. For example, instead of re-evaluating the Jacobian matrix using the latest iterate, one may always evaluate the Jacobian matrix at the initial guess. For a general mapping $\mathbf{G}(\mathbf{x})$: $\mathbb{R}^s \to \mathbb{R}^s$, Newton's method starts with an initial guess $\mathbf{x}^{(0)}$ and then defines the Newton sequence $\{\mathbf{x}^{(m)}\}$ from the solutions of the linear systems

$$\mathbf{G}'(\mathbf{x}^{(m-1)})(\mathbf{x}^{(m)} - \mathbf{x}^{(m-1)}) = -\mathbf{G}(\mathbf{x}^{(m-1)}), \qquad m = 1, 2, \ldots, \qquad (6.16)$$

where $\mathbf{G}'(\cdot)$ denotes the Jacobian matrix of $\mathbf{G}$. The fixed Jacobian or chord method again starts with some initial guess $\mathbf{x}^{(0)}$ and then generates the sequence $\{\mathbf{x}^{(m)}\}$ from the solutions of the linear systems

$$\mathbf{G}'(\mathbf{x}^{(0)})(\mathbf{x}^{(m)} - \mathbf{x}^{(m-1)}) = -\mathbf{G}(\mathbf{x}^{(m-1)}), \qquad m = 1, 2, \ldots, \qquad (6.17)$$

so that, unlike the situation for (6.16), for all m the same coefficient matrix is used. Thus, for example, $\mathbf{G}'(\mathbf{x}^{(0)})$ needs to be factored only once and then solving (6.17) for each value of m only requires the less expensive forward and back substitution procedures.

In the context of the discrete Navier–Stokes equations (6.1)–(6.2), the fixed Jacobian iteration (6.17) is described as follows. Given an initial guess $\mathbf{u}^{(0)} \in \mathbf{V}_0^h$, one generates the sequence $\{\mathbf{u}^{(m)}, p^{(m)}\}$ for $m = 1, 2, 3, \ldots$ by solving the sequence of *linear* problems

$$a(\mathbf{u}^{(m)}, \mathbf{v}^h) + c(\mathbf{u}^{(m)}, \mathbf{u}^{(0)}, \mathbf{v}^h) + c(\mathbf{u}^{(0)}, \mathbf{u}^{(m)}, \mathbf{v}^h) + b(\mathbf{v}^h, p^{(m)})$$

$$= (\mathbf{f}, \mathbf{v}^h) + c(\mathbf{u}^{(m-1)}, \mathbf{u}^{(0)}, \mathbf{v}^h) + c(\mathbf{u}^{(0)}, \mathbf{u}^{(m-1)}, \mathbf{v}^h)$$

$$- c(\mathbf{u}^{(m-1)}, \mathbf{u}^{(m-1)}, \mathbf{v}^h) \quad \text{for all } \mathbf{v}^h \in \mathbf{V}_0^h \qquad (6.18)$$

and

$$b(\mathbf{u}^{(m)}, q^h) = 0 \quad \text{for all } q^h \in S_0^h. \qquad (6.19)$$

Note that again the initial guess $\mathbf{u}^{(0)}$ is required to satisfy the boundary condition $\mathbf{u}^{(0)} = \mathbf{0}$ on Γ but is not required to satisfy the incompressibility constraint, even in its weak form (6.2). Also note that no initial guess for the pressure need be supplied.

In terms of the representations given in (6.5) the linear system (6.18)–(6.19) that determines $p^{(m)}$ and $\mathbf{u}^{(m)}$ is explicitly given by

$$\sum_{k=1}^{K} A_{\ell k} \beta_k^{(m)} + \sum_{j=1}^{J} B_{j\ell} \alpha_j^{(m)} = F_\ell^{(m)} \quad \text{for } \ell = 1, \ldots, K \qquad (6.20)$$

and

$$\sum_{k=1}^{K} B_{jk} \beta_k^{(m)} = 0 \quad \text{for } j = 1, \ldots, J, \qquad (6.21)$$

where

$$A_{\ell k} = a(\mathbf{v}_k, \mathbf{v}_\ell) + c(\mathbf{v}_k, \mathbf{u}^{(0)}, \mathbf{v}_\ell) + c(\mathbf{u}^{(0)}, \mathbf{v}_k, \mathbf{v}_\ell) \quad \text{for } k, \ell = 1, \ldots, K,$$

$$B_{jk} = b(\mathbf{v}_k, q_j) \quad \text{for } j = 1, \ldots, J \text{ and } k = 1, \ldots, K,$$

and

$$F_\ell^{(m)} = (\mathbf{f}, \mathbf{v}_\ell) + c(\mathbf{u}^{(m-1)}, \mathbf{u}^{(0)}, \mathbf{v}_\ell) + c(\mathbf{u}^{(0)}, \mathbf{u}^{(m-1)}, \mathbf{v}_\ell)$$

$$- c(\mathbf{u}^{(m-1)}, \mathbf{u}^{(m-1)}, \mathbf{v}_\ell) \quad \text{for } \ell = 1, \ldots, K.$$

The coefficient matrix of the linear system (6.20)–(6.21) clearly does not depend on m.

Unfortunately, this fixed Jacobian method is at best linearly convergent, and it requires "better" initial guesses in order to guarantee convergence (Gunzburger and Peterson [1983] and Karakashian [1982]).

For example, let (2.13) hold so that the solution of (6.1)–(6.2) is known
to be uniquely determined. Let d be defined by (6.11) and let the initial
guess satisfy

$$|\mathbf{u}^{(0)} - \mathbf{u}^h|_1 \leq \frac{d}{2} \quad \text{and} \quad \mathbf{u}^{(0)} = 0 \text{ on } \Gamma. \tag{6.22}$$

Then the sequence of iterates $\{\mathbf{u}^{(m)}, p^{(m)}\}$, $m = 1, 2, \ldots$ are uniquely
defined by (6.18)–(6.19), or equivalently, by (6.20)–(6.21) and
$|\mathbf{u}^{(m)} - \mathbf{u}^h|_1 \leq d/2$ for all m. Also, if $|\mathbf{u}^{(0)} - \mathbf{u}^h|_1 = d\varepsilon/2$ with $0 < \varepsilon < 1$,
then for $m \geq 1$

$$|\mathbf{u}^{(m)} - \mathbf{u}^h|_1 \leq \frac{d\varepsilon^{m+1}}{2} \tag{6.23}$$

and, for constant $C > 0$,

$$\|p^{(m)} - p^h\|_0 \leq C\varepsilon^{m+1}. \tag{6.24}$$

Thus one can guarantee convergence for initial guesses that are "twice
as good" as those used in Newton's method, i.e., compare (6.22) with
(6.12). Furthermore, (6.23) and (6.24) indicate that the convergence is
only linear.

In practical implementations of the method of (6.18)–(6.19), it is
found that the Jacobian matrix must be re-evaluated after every M
iterations, where M is a specified positive integer chosen empirically
in order to keep the iteration convergent. In this case, a linear system
with a new coefficient matrix must be solved every M iterations.

6.3.　A Simple Iteration Method

In case the uniqueness condition (2.13) holds, one may define an
iterative method that is globally convergent, i.e., converges for any
initial guess. For the discrete equations (6.1)–(6.2), this scheme is given
as follows. Given an initial guess $\mathbf{u}^{(0)}$, one generates the sequence
$\{\mathbf{u}^{(m)}, p^{(m)}\}$ for $m = 1, 2, 3, \ldots$ by solving the sequence of *linear* problems

$$a(\mathbf{u}^{(m)}, \mathbf{v}^h) + c(\mathbf{u}^{(m-1)}, \mathbf{u}^{(m)}, \mathbf{v}^h) + b(\mathbf{v}^h, p^{(m)})$$

$$= (\mathbf{f}, \mathbf{v}^h) \quad \text{for all } \mathbf{v}^h \in \mathbf{V}_0^h \tag{6.25}$$

and

$$b(\mathbf{u}^{(m)}, q^h) = 0 \quad \text{for all } q^h \in S_0^h. \tag{6.26}$$

The initial guess $\mathbf{u}^{(0)}$ is *not* required to satisfy the boundary condition $\mathbf{u}^{(0)} = \mathbf{0}$ on Γ nor the incompressibility constraint, even in its weak form (6.2). In fact, $\mathbf{u}^{(0)}$ need not belong to $\mathbf{V}_0^h$. Also note that again no initial guess for the pressure need be supplied.

In terms of the representations given in (6.5) the linear system (6.25)–(6.26) that determines $p^{(m)}$ and $\mathbf{u}^{(m)}$ is explicitly given by

$$\sum_{k=1}^{K} A_{\ell k}^{(m)} \beta_k^{(m)} + \sum_{j=1}^{J} B_{j\ell}\alpha_j^{(m)} = F_\ell^{(m)} \quad \text{for } \ell = 1, ..., K \qquad (6.27)$$

and

$$\sum_{k=1}^{K} B_{jk} \beta_k^{(m)} = 0 \quad \text{for } j = 1, ..., J, \qquad (6.28)$$

where

$$A_{\ell k}^{(m)} = a(\mathbf{v}_k, \mathbf{v}_\ell) + c(\mathbf{u}^{(m-1)}, \mathbf{v}_k, \mathbf{v}_\ell) \quad \text{for } k, \ell = 1, ..., K,$$

$$B_{jk} = b(\mathbf{v}_k, q_j) \quad \text{for } j = 1, ..., J \text{ and } k = 1, ..., K,$$

and

$$F_\ell^{(m)} = (\mathbf{f}, \mathbf{v}_\ell) \quad \text{for } \ell = 1, ..., K.$$

As was the case with Newton's method, a different linear system must be solved for at each step of the iteration. A precise convergence result for the method (6.25)–(6.26) is given as follows (Karakashian [1982]). Suppose that (2.13) is satisfied; then (6.1)–(6.2) have a unique solution and there exists a scalar α such that $0 < \alpha < 1$ and $\alpha v = \kappa_c |\mathbf{u}^h|_1$. Then the sequence $\{\mathbf{u}^{(m)}, p^{(m)}\}$ is uniquely defined by (6.25)–(6.26), or equivalently by (6.27)–(6.28). Moreover, for $m \geq 1$,

$$|\mathbf{u}^{(m)} - \mathbf{u}^h|_1 \leq \alpha^m |\mathbf{u}^{(0)} - \mathbf{u}^h|_1.$$

Thus whenever (2.13) is satisfied, the method (6.25)–(6.26) is *globally* and linearly convergent.

The global convergence of the method (6.25)–(6.26) and the rapid local convergence of Newton's method (6.3)–(6.4) may be taken advantage of by defining a composite method wherein one starts with a few iterations of the simple iterative scheme (6.25)–(6.26) and then one switches over to Newton's method in order to produce rapidly improving approximations to the solution of (6.1)–(6.2). The switch to Newton's method should occur when a simple iterate enters the attraction ball for Newton's method; unfortunately, one cannot know when this is the case. In practice one instead performs a fixed number of simple iterations, usually less than four, and then switches over to

Newton's method. This composite method may be easily implemented in an existing code using Newton's method since both its constitutive methods have the form

$$a(\mathbf{u}^{(m)}, \mathbf{v}^h) + \sigma c(\mathbf{u}^{(m)}, \mathbf{u}^{(m-1)}, \mathbf{v}^h) + c(\mathbf{u}^{(m-1)}, \mathbf{u}^{(m)}, \mathbf{v}^h) + b(\mathbf{v}^h, p^{(m)})$$

$$= (\mathbf{f}, \mathbf{v}^h) + \sigma c(\mathbf{u}^{(m-1)}, \mathbf{u}^{(m-1)}, \mathbf{v}^h) \quad \text{for all } \mathbf{v}^h \in \mathbf{V}_0^h \qquad (6.29)$$

and

$$b(\mathbf{u}^{(m)}, q^h) = 0 \quad \text{for all } q^h \in S_0^h, \qquad (6.30)$$

where $\sigma = 1$ for Newton's method (6.3)–(6.4) and $\sigma = 0$ for the simple iteration (6.25)–(6.26). Thus the composite scheme is defined as follows. Given an initial guess $\mathbf{u}^{(0)}$ one generates the sequence $\{\mathbf{u}^{(m)}, p^{(m)}\}$ for $m = 1, 2, 3, \ldots$ by solving the sequence of *linear* problems (6.29)–(6.30) where $\sigma = 0$ for $m \le M$ and $\sigma = 1$ for $m > M$ where M is the chosen number of simple iterations to be performed before one switches to Newton's method.

This composite method is an effective method for solving (6.1)–(6.2) whenever these have a unique solution, at least for two-dimensional problems. It remains of use even at higher values of the Reynolds number when used in conjunction with the continuation methods of Chapter 8.

6.4. Broyden's Method

Update, or quasi-Newton methods, are a class of methods for which the cost of generating each iterate is substantially less than that for Newton's method. For one thing, instead of solving a new linear system for every step of the iteration, low-rank updates of the inverse of an approximate Jacobian are effected. For another thing, many fewer function evaluations per step are needed, e.g., a considerably simpler assembly process is necessary. Due to the lack of symmetry and positive definiteness of the Jacobian matrix in the Navier–Stokes case, one may as well use the simplest update method, namely, Broyden's method. The use of Broyden's method for solving the discrete Navier–Stokes equations is studied in Engelman, Strang, and Bathe [1981], which we follow here.

In the context of a general mapping $\mathbf{G}(\mathbf{x})$: $\mathbb{R}^s \to \mathbb{R}^s$ the particular form of Broyden's method that we will use is described as follows.

One starts with an initial guess $\mathbf{x}^{(0)}$ for the solution of $\mathbf{G}(\mathbf{x}) = \mathbf{0}$ and an initial approximation H_0 for the inverse of the Jacobian matrix $\mathbf{G}'(\mathbf{x}^{(0)})$. For example, one may choose $H_0 = [\mathbf{G}'(\mathbf{x}^{(0)})]^{-1}$. Then the sequence $\{\mathbf{s}^{(m)}, \mathbf{x}^{(m)}, \mathbf{y}^{(m)}, H_m\}$ for $m = 1, 2, \ldots$ where $\{\mathbf{x}^{(m)}\}$ is the sequence of Broyden approximations to a solution of $\mathbf{G}(\mathbf{x}) = \mathbf{0}$, is defined by

$$\mathbf{s}^{(m)} = -H_{m-1}\mathbf{G}(\mathbf{x}^{(m-1)}), \tag{6.31}$$

$$\mathbf{x}^{(m)} = \mathbf{x}^{(m-1)} + \sigma_m \mathbf{s}^{(m)}, \tag{6.32}$$

$$\mathbf{y}^{(m)} = \mathbf{G}(\mathbf{x}^{(m)}) - \mathbf{G}(\mathbf{x}^{(m-1)}), \tag{6.33}$$

and

$$H_m = H_{m-1} + \frac{(\mathbf{s}^{(m)} - H_{m-1}\mathbf{y}^{(m)})\mathbf{s}^{(m)^T}}{\mathbf{s}^{(m)^T} H_{m-1}\mathbf{y}^{(m)}} H_{m-1}. \tag{6.34}$$

In (6.32) there are two options for choosing the scalar σ_m. One may simply choose $\sigma_m = 1$, in which case (6.31)–(6.34) describe a standard Broyden method; see Dennis and More [1977]. On the other hand, one may choose

$$\sigma_m = (\sigma \mid \min g_m(\sigma)) \quad \text{where } g_m(\sigma) = \mathbf{s}^{(m)^T}\mathbf{G}[\mathbf{x}^{(m)} + \sigma\mathbf{s}^{(m)}]. \tag{6.35}$$

Thus, σ_m is chosen to minimize $\mathbf{G}(\cdot)$ along the line $\mathbf{x}^{(m)} + \sigma\mathbf{s}^{(m)}$. The minimization (6.35) may be accomplished by a variety of optimization techniques for scalar functions of one variable; see Dennis and Schnabel [1983]. Note that any such minimization algorithm will require additional evaluations of the vector valued function $\mathbf{G}(\cdot)$ and thus it is not clear that such a sophisticated method for choosing σ_m yields a more efficient method than the simple choice $\sigma_m = 1$.

A serious drawback of Broyden's method, implemented following (6.31)–(6.34), is the need to store the approximate inverse Jacobian H_m. Note that even when $\mathbf{G}'(\mathbf{x}^{(0)})$ is a sparse matrix, e.g., a banded matrix, $H_0 = [\mathbf{G}'(\mathbf{x}^{(0)})]^{-1}$ and the subsequent H_m, $m \geq 1$, will be full matrices and storing even one of these is usually unacceptable in practical calculations. The following implementation (Engelman, Strang, and Bathe [1981] and Matthies and Strang [1977]) of Broyden's method avoids storing H_m. It requires the storage of some additional vectors as well as of the triangular factors of the initial Jacobian $\mathbf{G}'(\mathbf{x}^{(0)})$. One starts with some initial guess $\mathbf{x}^{(0)}$ and the initial Jacobian $\mathbf{G}'(\mathbf{x}^{(0)})$. The first step is essentially a Newton step, i.e., we solve $\mathbf{G}'(\mathbf{x}^{(0)})\mathbf{s}^{(1)} = -\mathbf{G}(\mathbf{x}^{(0)})$ for the first search direction $\mathbf{s}^{(1)}$ and then set

$\mathbf{x}^{(1)} = \mathbf{x}^{(0)} + \sigma_1 \mathbf{s}^{(1)}$ where $\sigma_1 = 1$ or is found through the minimization (6.35). We also store the triangular factors of $\mathbf{G}'(\mathbf{x}^{(0)})$. Then, for $m = 2, 3, \ldots$

Solve $\mathbf{G}'(\mathbf{x}^{(0)})\mathbf{q}^{(1)} = -\mathbf{G}(\mathbf{x}^{(m-1)})$ for $\mathbf{q}^{(1)}$.

For $j = 1, \ldots, m - 2$ set $\mathbf{q}^{(j+1)} = \mathbf{q}^{(j)} + \rho_j(\mathbf{t}^{(j)} - \mathbf{r}^{(j)})\mathbf{t}^{(j)^T}\mathbf{q}^{(j)}$,

set $\mathbf{r}^{(m)} = \mathbf{q}^{(m-1)} - \mathbf{s}^{(m-1)}$,

$$\mathbf{t}^{(m)} = \sigma_{m-1}\mathbf{s}^{(m-1)},$$

$$\rho_m = \frac{1}{\mathbf{t}^{(m)^T}\mathbf{r}^{(m)}},$$

$$\mathbf{s}^{(m)} = \mathbf{q}^{(m-1)} + \rho_m(\mathbf{t}^{(m)} - \mathbf{r}^{(m)})\mathbf{t}^{(m)^T}\mathbf{q}^{(m-1)}$$

and

$$\mathbf{x}^{(m)} = \mathbf{x}^{(m-1)} + \sigma_m\mathbf{s}^{(m)}.$$

We see that the vectors $\mathbf{x}^{(m)}$, $\mathbf{s}^{(m)}$, $\mathbf{r}^{(m)}$, and $\mathbf{t}^{(m)}$ must be stored, as well as the factors of $\mathbf{G}'(\mathbf{x}^{(0)})$ and the scalars ρ_m and σ_m, the latter either set to unity or found through the minimization (6.35). For large values of m this storage cost becomes rather onerous. Thus, in practice, the algorithm should be restarted every M steps, i.e., we set $\mathbf{x}^{(0)} = \mathbf{x}^{(M)}$, compute and factor $\mathbf{G}'(\mathbf{x}^{(0)})$, etc., through another M steps. If a good initial guess is available, then such a restarting procedure will probably be invoked at most once if one chooses M between 5 and 10 (Engleman, Strang, and Bathe [1981]).

In the context of the discrete Navier–Stokes equations (6.1)–(6.2), we set $\mathbf{x}^{(m)} = (\alpha_1^{(m)}, \ldots, \alpha_J^{(m)}, \beta_1^{(m)}, \ldots, \beta_K^{(m)})^T$ where the $\alpha_j^{(m)}$'s and $\beta_k^{(m)}$'s are the coefficients in the representation (6.5). Of course, this numbering system for the unknowns would not be used in practice; rather, the velocity and pressure unknowns would be interspersed so as to, e.g., minimize the bandwidth. The components of $\mathbf{G}(\mathbf{x})$ are defined by (2.3)–(2.4), i.e.,

$$G_\ell[\alpha_1^{(m)}, \ldots, \alpha_J^{(m)}, \beta_1^{(m)}, \ldots, \beta_K^{(m)}] = \sum_{k=1}^{K} a(\mathbf{v}_\ell, \mathbf{v}_k)\beta_k^{(m)}$$

$$+ \sum_{k,j=1}^{K} c(\mathbf{v}_j, \mathbf{v}_k, \mathbf{v}_\ell)\beta_k^{(m)}\beta_j^{(m)} + \sum_{j=1}^{J} b(\mathbf{v}_\ell, q_j)\alpha_j^{(m)} - (\mathbf{f}, \mathbf{v}_\ell)$$

for $\ell = 1, \ldots, K$

and

$$G_\ell[\alpha_1^{(m)}, \dots, \alpha_J^{(m)}, \beta_1^{(m)}, \dots, \beta_K^{(m)}] = \sum_{k=1}^{K} b(\mathbf{v}_k, q_{\ell-K})\beta_k^{(m)}$$

for $\ell = K + 1, \dots, K + J$.

Here the $\mathbf{v}_j$'s and q_ℓ's are the basis functions for the velocity and pressure finite element spaces, respectively. Again, in practice, the equations would not be numbered in this manner, but would be reordered so as to, e.g., minimize the bandwidth. The initial Jacobian $J_0 = \mathbf{G}'[\mathbf{x}^{(0)}]$ is explicitly given by

$$(J_0)_{\ell k} = A_{\ell k}^{(0)} \quad \text{for } \ell, k = 1, \dots, K,$$

$$(J_0)_{\ell k} = (J_0)_{k\ell} = B_{k, \ell-K} \quad \text{for } \ell = K + 1, \dots, K + J \text{ and } k = 1, \dots, K$$

and

$$(J_0)_{\ell k} = 0 \quad \text{for } \ell, k = K + 1, \dots, K + J,$$

where $A_{\ell k}^{(0)}$ and B_{jk} are defined by (6.8) and (6.9), respectively.

Concerning the convergence of the Broyden iterates, it is known that they will converge to a nonsingular solution of (6.1)–(6.2) provided the initial guess is sufficiently close to that solution. No quantitative information is known about how good the initial guess need be; however, when the method converges, it is known to be superlinearly convergent, i.e., not only does $|\mathbf{u}^{(m)} - \mathbf{u}^h|_1 \to 0$ as $m \to \infty$, but

$$\lim_{m \to \infty} \frac{|\mathbf{u}^{(m+1)} - \mathbf{u}^h|_1}{|\mathbf{u}^{(m)} - \mathbf{u}^h|_1} = 0.$$

The convergence performance of Broyden's method, though not in general as good as the quadratic convergence of Newton's method, is superior to that of the linearly convergent methods of Sections 6.2 and 6.3; indeed, in practice, superlinear convergence is acceptable.

6.5. A Solution Method Using an Equivalent Optimization Problem

One may restate the discretized Navier–Stokes equations in the form of a minimization problem whose solution can be found by solving a sequence of linear Stokes problems. Here, we only consider the simple case of stationary flows with homogeneous velocity boundary conditions. Our treatment is derived from Bristeau *et al.* [1979, 1980a, and 1980b], Glowinski [1984] and Glowinski, and Periaux and Pironneau [1980], which may be consulted for details and for a discussion of this approach in more general settings.

Given a div-stable pair of finite element spaces $\mathbf{V}_0^h$ and S_0^h, consider the functional

$$J(\mathbf{w}^h) = \tfrac{1}{2}a\,[\xi^h(\mathbf{w}^h), \xi^h(\mathbf{w}^h)] \quad \text{for all } \mathbf{w}^h \in \mathbf{V}_0^h, \qquad (6.36)$$

where $\xi^h \in \mathbf{V}_0^h$ is related to $\mathbf{w}^h \in \mathbf{V}_0^h$ through the discrete Stokes problem,

$$a(\xi^h, \mathbf{v}^h) + b(\mathbf{v}^h, \sigma^h) = a(\mathbf{w}^h, \mathbf{v}^h) + c(\mathbf{w}^h, \mathbf{w}^h, \mathbf{v}^h) - (\mathbf{f}, \mathbf{v}^h)$$

$$\text{for all } \mathbf{v}^h \in \mathbf{V}_0^h, \qquad (6.37)$$

and

$$b(\xi^h, q^h) = 0 \quad \text{for all } q^h \in S_0^h, \qquad (6.38)$$

and where σ^h is an auxiliary pressure type variable. Now, seek $\mathbf{u}^h \in \mathbf{V}_0^h$ such that

$$J(\mathbf{u}^h) \leq J(\mathbf{w}^h) \quad \text{for all } \mathbf{w}^h \in \mathbf{V}_0^h. \qquad (6.39)$$

The bilinear form $a(\cdot, \cdot)$ is symmetric and positive definite on $\mathbf{V}_0^h \times \mathbf{V}_0^h$, so that it easily follows that a minimizer $\mathbf{u}^h \in \mathbf{Z}^h$ of $J(\cdot)$, if such a minimizer exists, is a solution of the discrete Navier–Stokes equations (2.1)–(2.2); conversely, if $\mathbf{u}^h$ is a solution of (2.1)–(2.2), it is a minimizer of $J(\cdot)$ over $\mathbf{V}_0^h$. To see this, first note that if a $\mathbf{u}^h \in \mathbf{Z}^h$ can be found such that $J(\mathbf{u}^h) = 0$, then clearly this $\mathbf{u}^h$ minimizes $J(\cdot)$ over $\mathbf{V}_0^h$. But $J(\mathbf{u}^h) = 0$ implies, from its definition (6.36), that $\xi^h(\mathbf{u}^h) = 0$. Then, from (6.37)–(6.38) we conclude that the pair $(\mathbf{u}^h, -\sigma^h)$ is a solution of (2.1)–(2.2); in fact, $\sigma^h = -p^h$. Thus, if a minimizer $\mathbf{u}^h \in \mathbf{Z}^h$ exists, it is a solution of (2.1)–(2.2). On the other hand, if $\mathbf{u}^h$ is a solution of (2.1)–(2.2), then setting $\mathbf{w}^h = \mathbf{u}^h \in \mathbf{Z}^h$ and letting $\mathbf{v}^h \in \mathbf{Z}^h$ be arbitrary in (6.37) yields that (6.37)–(6.38) is a homogeneous Stokes problem whose unique solution is $\xi^h(\mathbf{u}^h) = 0$; thus, the candidate $\mathbf{u}^h$ renders $J(\mathbf{u}^h) = 0$, i.e., it minimizes $J(\cdot)$. Thus, we have shown that solutions of (2.1)–(2.2) are minimizers, and thus a minimizer belonging to $\mathbf{Z}^h$ exists. The task at hand is then to find a minimizer of $J(\cdot)$ belonging to $\mathbf{Z}^h$.

Now that we have established the correspondence between solutions of the problems (2.1)–(2.2) and (6.36)–(6.39), we proceed to show how the latter may be solved. The method is initialized by picking a guess $\mathbf{u}^{(0)} \in \mathbf{Z}^h$. Then, at the start of the kth step of the iteration, we assume we know $\mathbf{u}^{(k)} \in \mathbf{Z}^h$. We then solve the linear Stokes problem

$$a(\xi^{(k)}, \mathbf{v}^h) + b(\mathbf{v}^h, \sigma^{(k)}) = a(\mathbf{u}^{(k)}, \mathbf{v}^h) + c(\mathbf{u}^{(k)}, \mathbf{u}^{(k)} \mathbf{v}^h) - (\mathbf{f}, \mathbf{v}^h)$$

$$\text{for all } \mathbf{v}^h \in \mathbf{V}_0^h \tag{6.40}$$

and

$$b(\xi^{(k)}, q^h) = 0 \qquad \text{for all } q^h \in S_0^h \tag{6.41}$$

for $\xi^{(k)} \in \mathbf{Z}^h$ and $\sigma^{(k)} \in S_0^h$. We then solve another Stokes problem

$$a(\eta^{(k)}, \mathbf{v}^h) + b(\mathbf{v}^h, \theta^{(k)}) = a(\xi^{(k)}, \mathbf{v}^h) + c(\mathbf{v}^h, \mathbf{u}^{(k)}, \xi^{(k)})$$

$$+ c(\mathbf{u}^{(k)}, \mathbf{v}^h, \xi^{(k)}) \qquad \text{for all } \mathbf{v}^h \in \mathbf{V}_0^h \tag{6.42}$$

and

$$b(\eta^{(k)}, q^h) = 0 \qquad \text{for all } q^h \in S_0^h \tag{6.43}$$

for $\eta^{(k)} \in \mathbf{Z}^h$ and $\theta^{(k)} \in S_0^h$. Next, if $k \geq 1$, set

$$\alpha_k = \frac{a(\eta^{(k)} - \eta^{(k-1)}, \eta^{(k)})}{a(\eta^{(k)}, \eta^{(k)})}. \tag{6.44}$$

and

$$\zeta^{(k)} = \eta^{(k)} + \alpha_k \zeta^{(k-1)}. \tag{6.45}$$

Otherwise, i.e., if $k = 0$, set $\zeta^{(0)} = \eta^{(0)}$. Clearly $\zeta^{(k)} \in \mathbf{Z}^h$. The next step is to solve the one-dimensional minimization

$$\text{Find } \lambda_k = \{\lambda \in \mathbb{R} : J(\mathbf{u}^{(k)} - \lambda\zeta^{(k)}) \text{ is a minimum}\}. \tag{6.46}$$

Finally, set

$$\mathbf{u}^{(k+1)} = \mathbf{u}^{(k)} - \lambda_k \zeta^{(k)}. \tag{6.47}$$

Clearly, if $\mathbf{u}^{(0)} \in \mathbf{Z}^h$ then all the subsequent iterates $\mathbf{u}^{(k)} \in \mathbf{Z}^h$ as well. If worse comes to worse, such an initial guess can be generated from an arbitrary initial guess $\mathbf{u}^{(-1)} \notin \mathbf{Z}^h$ or even $\mathbf{V}_0^h$ by solving the initial Stokes problem

$$a(\mathbf{u}^{(0)}, \mathbf{v}^h) + b(\mathbf{v}^h, p^{(0)}) = a(\mathbf{u}^{(-1)}, \mathbf{v}^h) \qquad \text{for all } \mathbf{v}^h \in \mathbf{V}_0^h$$

$$b(\mathbf{u}^{(0)}, q^h) = 0 \qquad \text{for all } q^h \in S_0^h.$$

Very often we may simply pick $\mathbf{u}^{(0)} = \mathbf{0}$.

The determination of $\mathbf{u}^{(k+1)}$ from $\mathbf{u}^{(k)}$ requires the solution of a series of discrete Stokes problems. First, we have the discrete Stokes problem

(6.40)–(6.41) to determine $\xi^{(k)}$; next, we have the discrete Stokes problem (6.42)–(6.43) to determine $\eta^{(k)}$. Also, the search for the scalar λ_k in (6.46) will surely require a few evaluations of the functional $J(\cdot)$; each of these evaluations requires the solution of a discrete Stokes problem. As is the case for many other algorithms for solving the discrete Navier–Stokes equations, the availability of efficient solvers for the discrete Stokes equation is a necessary adjunct to the method of (6.40)–(6.47). There is ample discussion of such solvers available in the literature; see, e.g., Glowinski [1984], Temam [1979], and Thomasset [1981] and some of the discussion of Section 7.1.

7

Solving the Linear Systems

Any of the iterative methods of Chapter 6 requires the solution of at least one linear system of equations. For many two-dimensional problems this may be often accomplished by a direct banded elimination procedure. For the linear Stokes equations some iterative methods have also been developed, e.g., multigrid methods (Verfurth [1984b and 1988]) and conjugate gradient methods (Glowinski [1984]). These are also useful for the nonlinear Navier–Stokes equations whenever these are solved through an iterative method that requires the solution of a sequence of Stokes solvers such as the method of Section 6.5. For an application of multigrid methods to the solution of the discretized Navier–Stokes equations, see Ghia, Ghia, and Shin [1982].

Here we briefly mention two methods for reducing the number of unknowns in the linear systems resulting from linearizations of (6.1)–(6.2) effected within some iterative method. We then discuss a particular domain decomposition method, i.e., substructuring, which is useful for solving these linear systems, especially in the environment of computers with parallel processor architectures.

Before continuing, we remark on two issues concerning the direct solution of linearizations of (6.1)–(6.2). The first is whether or not pivoting is necessary when the linear systems in question here are solved by an elimination procedure such as banded elimination. The coefficient matrices encountered are nonsymmetric and indefinite, excepting for matrices resulting from discretizations of the Stokes problem that are symmetric and indefinite. Moreover, due to the fact that the pressure unknown does not appear in the continuity equation, these matrices will have many zero diagonal entries. Therefore, in general, it seems that pivoting must be invoked in order for any Gaussian elimination method to proceed in a stable manner. However, in practice, it is often the case that one may order the unknowns so that pivoting may be safely avoided. The general rule of thumb is that pressure degrees of freedom associated with a particular node or element should be numbered *after* the velocity degrees of freedom corresponding to the same node or element.

Although for two-dimensional problems direct solution techniques have proven to be popular and successful, it seems that they cannot be used for solving realistic three-dimensional problems on today's super-computers. In fact, the most efficient possible implementation of the Gauss elimination method is via nested dissection techniques (George and Liu [1981] and Nicolaides and Wu [1988]); however, even when such an implementation is possible, the storage and work requirements are beyond the capacity of present hardware; see Nicolaides and Wu [1988].

7.1. Reducing the Number of Unknowns

The linear systems encountered in Chapter 6 have the form

$$\begin{pmatrix} A & B^T \\ B & 0 \end{pmatrix}\begin{pmatrix} \mathbf{U} \\ \mathbf{P} \end{pmatrix} = \begin{pmatrix} \mathbf{F} \\ \mathbf{G} \end{pmatrix}, \tag{7.1}$$

where A and B are matrices, the vectors $\mathbf{U}$ and $\mathbf{P}$ contain the unknown velocity and pressure degrees of freedom, respectively, and the vectors $\mathbf{F}$ and $\mathbf{G}$ result from inhomogeneous data. In particular, for the boundary condition (1.3), $\mathbf{G} = \mathbf{0}$. For an example of (7.1), examine (6.6)–(6.7).

One way to reduce the number of unknowns in (7.1) is to eliminate the pressure. To this end consider, for $\ell = 0, 1, 2, \ldots$, the sequence of systems

$$\begin{pmatrix} A & B^T \\ B & -\varepsilon M \end{pmatrix}\begin{pmatrix} \mathbf{U}^{(\ell+1)} \\ \mathbf{P}^{(\ell+1)} \end{pmatrix} = \begin{pmatrix} \mathbf{F} \\ \mathbf{G} - \varepsilon M \mathbf{P}^{(\ell)} \end{pmatrix} \tag{7.2}$$

where $\mathbf{P}^{(0)}$ is a suitably chosen initial guess, e.g., $\mathbf{P}^{(0)} = \mathbf{0}$, and M is a suitably chosen matrix, e.g., usually $M = I$. From (7.2) one easily finds

$$\left(A + \frac{1}{\varepsilon}B^T M^{-1} B\right)\mathbf{U}^{(0)} = \mathbf{F} + \frac{1}{\varepsilon}B^T M^{-1}\mathbf{G} - B^T\mathbf{P}^{(0)} \qquad (7.3)$$

and

$$\left(A + \frac{1}{\varepsilon}B^T M^{-1} B\right)\mathbf{U}^{(\ell+1)} = A\mathbf{U}^{(\ell)} + \frac{1}{\varepsilon}B^T M^{-1}\mathbf{G} \qquad \text{for } \ell = 1, 2, \dots . \quad (7.4)$$

This method is the matrix analogue of the iterated penalty, or augmented Lagrangian, method mentioned in Section 5.2. See Glowinski [1984], Gunzburger, Liu, and Nicolaides [1983] and Segal [1979]. It can be shown, for sufficiently small values of the Reynolds number, or for a vicinity of a branch of a nonsingular solution, that the sequence $\{\mathbf{U}^{(\ell)}\}$ defined by (7.3)–(7.4) converges linearly to the solution $\mathbf{U}$ of (7.1), i.e., in the ℓ_2-norm

$$\|\mathbf{U} - \mathbf{U}^{(\ell+1)}\| \le C\varepsilon\|\mathbf{U} - \mathbf{U}^{(\ell)}\|.$$

It should be noted that (7.2), or equivalently (7.3)–(7.4), is not a penalty method; indeed, if $\mathbf{P}^{(0)} = \mathbf{0}$ and $M = I$, then $\mathbf{U}^{(1)}$ would be the result of a standard penalty method. In this case $\|\mathbf{U} - \mathbf{U}^{(1)}\| = O(\varepsilon)$. Thus, by continuing the iteration (7.2), an arbitrarily good solution of (7.1) may be obtained. Since in practice $\varepsilon = 0.001$ or 0.0001 are reasonable choices, one should never have to perform more than two steps of (7.2). We note the obvious facts that the systems in (7.3)–(7.4) have less unknowns than does the system (7.1) and that all of the systems in (7.3)–(7.4) involve the same coefficient matrix, which needs to be factored only once. The latter implies that the cost of carrying out more than one iteration is not much greater than the cost of a single iteration.

A second method, applicable to the case $\mathbf{G} = \mathbf{0}$, for reducing the number of unknowns in (7.1) is to find a basis for the null space of the discrete divergence matrix B. Suppose the matrix Q has columns that form a basis for the right null space of B. Then, since $B\mathbf{U} = 0$, we have that $\mathbf{U} = Q\mathbf{V}$ for some vector $\mathbf{V}$; then, from (7.1)

$$(Q^T A Q)\mathbf{V} = Q^T\mathbf{F}. \qquad (7.5)$$

Once (7.5) is solved for $\mathbf{V}$, the discrete velocity is recovered from $\mathbf{U} = Q\mathbf{V}$. The advantage of (7.5) over (7.1) is that the former involves many fewer unknowns, i.e., the number of components in **V** is roughly the number of components in **U** *less* the number of components in **P**. Of course, in order to form (7.5), the basis matrix Q must be found. For piecewise bilinear velocity finite element spaces based on uniform "triangulations" of the flow domain, this may be efficiently accomplished using graph theoretic arguments (Gustafson and Hartman [1983] and Hall *et al.* [1985]. For the more general case, see Berry *et al.* [1985]; for higher order elements, see Scott and Vogelius [1985a].

It should be noted that versions of both the above algorithms that may be directly applied to the nonlinear system (6.1)–(6.2) can also be easily defined. See the cited references for details.

In case the system of equations (7.1) represents a discrete Stokes problem, we have the added advantage that the matrix A is symmetric and positive definite. This implies that the coefficient matrices in (7.3), (7.4), and (7.5) are also symmetric and positive definite, so that these problems may be solved by iterative techniques such as the conjugate gradient method. See Glowinski [1984] for other methods that take advantage of this feature of the discrete Stokes equations.

7.2. A Substructuring Method

Substructuring algorithms are in general use for the efficient solution of the discrete equations arising from discretizations of structural mechanics problems. Typically, these problems give rise to positive definite linear systems. Substructuring algorithms are not often used for fluid mechanics problems because in this case one encounters indefinite and nonsymmetric matrices. Thus any associated block solution procedure, in a straightforward implementation, requires pivoting, a situation that negates the advantages of the substructuring algorithm. Here we discuss a substructuring method for the discrete Navier–Stokes equations that does not require pivoting. The method is a special case of a method of Gunzburger and Nicolaides [1985] that is applicable to general linear systems. Details concerning the algorithm in the Navier–Stokes context, especially with regard to its efficient implementation, may be found in Gunzburger and Nicolaides [1984 and 1986].

We consider the solution of the linear systems of the form (7.1) that arise when attempting to solve the discrete Navier–Stokes equations

(6.1)–(6.2). "Natural" orderings of the unknowns appearing in these linear systems, e.g., numbering left to right and then bottom to top, typically result in banded linear systems. The substructuring method results from a different choice for the numbering of the unknowns that is based on a decomposition of the domain Ω into subdomains.

We subdivide the region Ω into open subregions Ω_i, $i = 1, \ldots, m$, such that $\bar\Omega = \bigcup_{i=1}^{m} \bar\Omega_i$ and $\Omega_i \cap \Omega_j = 0$ for $i \neq j$. We denote by Γ_{ij}, $i, j = 1, \ldots, m$, the interfaces between these subregions, i.e., $\Gamma_{ij} = \bar\Omega_i \cap \bar\Omega_j$. Of course, $\Gamma_{ij} = \Gamma_{ji}$ and for particular values of i and j in a given subdivision, Γ_{ij} may be empty. See Figure 7.1 for an illustration. We assume that the interfaces Γ_{ij} coincide with faces or edges of finite elements belonging to the finer subdivision of Ω, which is used to define the finite element spaces used in (6.1)–(6.2). When the pressure and velocity triangulations differ, this assumption is applied to the coarser pressure triangulation. It is tacitly assumed that the subdivision of Ω into finite elements is substantially finer than that into the subregions Ω_i, $i = 1, \ldots, m$.

All unknowns and equations associated with the interior of a subdomain Ω_i are grouped together, one subdomain at a time, and unknowns and equations associated with the interfaces Γ_{ij} are grouped together and numbered last. For example, some unknowns, i.e., trial functions, and equations, i.e., test functions, are associated with vertices or edges of triangles. Those corresponding to the interior of

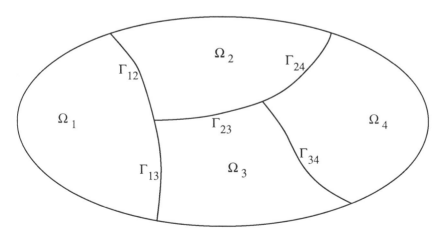

Figure 7.1
A subdivision of a domain into four subdomains.

Ω_1 would be numbered first, then proceeding to Ω_2, etc., and finally to Ω_m. Then one would number all the unknowns corresponding to vertices and edges that lie on the interfaces Γ_{ij}, $i,j = 1, \ldots, m$. In addition, there may be test and trial functions more naturally associated with the interior of the finite elements, e.g., bubble functions, and the unknowns and equations corresponding to these functions are grouped together with the other ones associated with the interior of the corresponding subregion. Note that discontinuous pressure test and trial functions should be associated with the interior of the elements, since any association with the boundary would result in an ambiguous definition of these functions along the interfaces.

The net result of the above numbering scheme is that the linear systems arising from the iterative methods of Chapter 6 have the form

$$
\begin{pmatrix}
A_1 & & & & B_1 \\
 & A_2 & & & B_2 \\
 & & \ddots & & \vdots \\
 & & & A_m & B_m \\
C_1 & C_2 & \cdots & C_m & A_I
\end{pmatrix}
\begin{pmatrix}
\mathbf{W}_1 \\
\mathbf{W}_2 \\
\vdots \\
\mathbf{W}_m \\
\mathbf{W}_I
\end{pmatrix}
=
\begin{pmatrix}
\mathbf{F}_1 \\
\mathbf{F}_2 \\
\vdots \\
\mathbf{F}_m \\
\mathbf{F}_I
\end{pmatrix}.
\tag{7.6}
$$

In (7.6) the matrices A_i, $i = 1, \ldots, m$, result in the case of both the test and trial functions being associated with the interior of the subregions Ω_i, $i = 1, \ldots, m$, respectively. The matrix A_I arises when both the test and trial functions are associated with the interfaces Γ_{ij}, $i,j = 1, \ldots, m$. The matrices C_i (B_i) arise from trial (test) functions associated with the interiors of the subdomains Ω_i and test (trial) functions associated with the interfaces Γ_{ij}. The vectors $\mathbf{W}_i$, $i = 1, \ldots, m$, respectively, denote unknowns, i.e., velocity and pressure degrees of freedom, associated with the interiors of Ω_i, while $\mathbf{W}_I$ denotes unknowns associated with the interfaces. A similar partitioning of equations can be associated with the right-hand side vectors $\mathbf{F}_i$, $i = 1, \ldots, m$, and $\mathbf{F}_I$.

Should the matrices A_i be invertible, then an obvious block elimination procedure, without interchanges, may be used to efficiently solve (7.6). Indeed, for positive definite problems the matrices A_i themselves are positive definite. However, for discretizations of the Navier–Stokes equations that involve only discontinuous pressure test and trial functions, the matrices A_i are singular. Indeed, these matrices are exactly those that arise from the analogous discretization of the problems

$$-v\,\Delta\mathbf{u} + \mathbf{u} \cdot \operatorname{grad} \mathbf{u} + \operatorname{grad} p = \mathbf{f} \text{ in } \Omega_i$$

$$\operatorname{div} \mathbf{u} = 0 \text{ in } \Omega_i \quad \text{and} \quad \mathbf{u} = \mathbf{0} \text{ on } \Gamma_i,$$

where Γ_i, the boundary of Ω_i, may consist of part of Γ as well as some of the interfaces Γ_{ij}. Indeed, $\Gamma_i = (\bar{\Omega}_i \cap \Gamma) \cup (\bigcup_{j=1}^{m} \Gamma_{ij})$. Thus each of the matrices A_i has a single local pressure null vector, i.e., the pressure function that is constant over Ω_i. Since in this case the matrices A_i are singular, a standard block elimination procedure cannot be used. On the other hand, if one uses continuous discrete pressure test and trial functions, then some of the interface unknowns within the vector $\mathbf{W}_I$ correspond to pressure degrees of freedom and the matrices A_i are non-singular.

It is the case that usually one does not *a priori* impose the zero mean condition on the pressure space. This is because this condition is global in nature, i.e., it involves *all* the pressures unknowns, and is thus cumbersome to implement. Then (7.6) determines the pressure only up to an additive constant, over Ω, function. On the other hand, (7.6) *always* uniquely determines the velocity degrees of freedom.

We first consider the case where the matrices A_i are nonsingular, i.e., continuous pressure finite element spaces. In this case one easily finds that

$$D\mathbf{W}_I = \mathbf{G} \tag{7.7}$$

and

$$\mathbf{W}_i = A_i^{-1}(\mathbf{F}_i - B_i\mathbf{W}_I) \quad \text{for } i = 1, \ldots, m, \tag{7.8}$$

where

$$D = A_I - \sum_{i=1}^{m} C_i A_i^{-1} B_i \quad \text{and} \quad \mathbf{G} = \mathbf{F}_I - \sum_{i=1}^{m} C_i A_i^{-1} \mathbf{F}_i. \tag{7.9}$$

Thus the solution of (7.6) is found by forming the matrix D and vector $\mathbf{G}$ according to (7.9), then solving for $\mathbf{W}_I$ from (7.7), and then solving for $\mathbf{W}_i$, $i = 1, \ldots, m$, from (7.8). Of course, in (7.8)–(7.9) the matrix inverses are not explicitly calculated, but rather appropriate linear systems are solved. If the zero mean over Ω condition is not imposed on the discrete pressure space, then the matrix D will be singular. However, (7.7) is still easily solved since the null space of D is explicitly known, i.e., it is given by setting all velocity degrees of freedom in $\mathbf{W}_I$ to zero and all pressure degrees of freedom in that vector to the same constant value.

In case one employs only discontinuous pressure test and trial functions, a slightly more complicated version of the algorithm (7.7)–(7.9) needs to be used to account for the singularity of the matrices A_i. First we note that in this case we have that, for $i = 1, \ldots, m$,

$$\mathbf{W}_i = \mathbf{Y}_i + \sigma_i \mathbf{N}_i, \tag{7.10}$$

where $A_i \mathbf{N}_i = 0$ and $\mathbf{N}_i^T \mathbf{Y}_i = 0$. The null vector $\mathbf{N}_i$ is explicitly known; the entries of this vector corresponding to velocity degrees of freedom are zero and those corresponding to pressure degrees of freedom are unity. The scalars σ_i and the vectors $\mathbf{Y}_i$, $i = 1, \ldots, m$, are at this point not determined. From (7.6) one easily finds that

$$D\mathbf{W}_I = \mathbf{G} - \sum_{i=1}^{m} \sigma_i C_i \mathbf{N}_i \tag{7.11}$$

and

$$\mathbf{Y}_i = A_i^+ (\mathbf{F}_i - B_i \mathbf{W}_I) \quad \text{for } i = 1, \ldots, m, \tag{7.12}$$

where

$$D = A_I - \sum_{i=1}^{m} C_i A_i^+ B_i \quad \text{and } \mathbf{G} = \mathbf{F}_I - \sum_{i=1}^{m} C_i A_i^+ \mathbf{F}_i \tag{7.13}$$

and where A_i^+ denotes the pseudo-inverse of A_i (Golub and van Loan [1983]). Since in the case under consideration the vector $\mathbf{W}_I$ consists exclusively of velocity degrees of freedom, and these are uniquely determined, the matrix D appearing in (7.11) and defined in (7.13) is nonsingular. Likewise, because of the properties of the pseudo-inverse, (7.12) uniquely determines $\mathbf{Y}_i$ from $\mathbf{W}_I$. Therefore all that remains to define the solution of (7.6) through (7.10)–(7.13) is to find the constants σ_i. It can be shown that these may be determined from the matrix problem

$$RS = H, \tag{7.14}$$

where $\mathbf{S} = (\sigma_1, \sigma_2, \ldots, \sigma_m)^T$, $\mathbf{H} = (\mathbf{T}_1^T, \mathbf{T}_2^T, \ldots, \mathbf{T}_m^T)^T$,

$$R = \begin{pmatrix} \mathbf{R}_{11} & \cdots & \mathbf{R}_{1m} \\ \vdots & & \vdots \\ \mathbf{R}_{m1} & \cdots & \mathbf{R}_{mm} \end{pmatrix}$$

$$\mathbf{T}_i = (I - A_i A_i^+)(\mathbf{F}_i - B_i D^{-1} \mathbf{G}) \quad \text{for } i = 1, \ldots, m,$$

and

$$\mathbf{R}_{ij} = (I - A_i A_i^+) B_i D^{-1} C_j \mathbf{N}_j \quad \text{for } i, j = 1, \ldots, m.$$

The matrix R is rectangular with m columns and with a number of rows equal to the sum of the number of rows of the matrices A_i, $i = 1, ..., m$. However, the linear system (7.14) is consistent and its solution may be found by solving the $m \times m$ system

$$R^T R \mathbf{S} = R^T \mathbf{H}. \qquad (7.15)$$

Now, if the zero mean condition on the pressure has been imposed a priori on the pressure trial and test functions, then $R^T R$ is nonsingular and (7.15) may be solved for $\mathbf{S}$. On the other hand, if this constraint is not imposed, then $R^T R$ has a one-dimensional null space consisting of the vector all of whose components have the same constant value. In any case, the small problem (7.15) is easily solved.

To summarize, in order to solve (7.6) when one only uses discontinuous discrete pressure test and trial functions, one first sets up and solves (7.15) for $\mathbf{S}$. One then solves (7.11) for $\mathbf{W}_I$ and computes $\mathbf{Y}_i$, $i = 1, ..., m$, from (7.12). Finally, $\mathbf{W}_i$, $i = 1, ..., m$, is computed from (7.10). One does not have to explicitly calculate the pseudo-inverses A_i^+; for a given vector $\mathbf{X}$, $A_i^+ \mathbf{X}$ may be computed by, e.g., an elimination procedure, in much the same way as $A_i^{-1} \mathbf{X}$ can be computed whenever A_i is invertible. For details of such an implementation, see Gunzburger and Nicolaides [1984 and 1986]. These references may also be consulted for a discussion of the inherent parallelism in the algorithms of this section as well as for issues concerning their efficient implementation. Of course, one obvious conclusion is that substructuring algorithms are considerably easier to apply to discretizations using continuous pressure finite element test and trial functions.

8

Solution Methods for Large Reynolds Numbers

8.1. Continuation Methods

The methods of Chapter 6, or for that matter, almost all methods for solving the nonlinear system of discrete equations (6.1)–(6.2), being iterative in character, require "good" initial guesses. For low values of the Reynolds numbers, the method of Section 6.3 is known to be globally and linearly convergent, i.e., it produces a sequence of iterates that, starting with any initial guess for the discrete velocity field, converges linearly to the solution of the discrete equations (6.1)–(6.2). If faster convergence is desired, then this method may be used in conjunction with another method, e.g., Newton's method. Thus one way to generate a good initial guess for Newton's or Broyden's method is to use the simple iteration scheme of Section 6.3.

For higher values of the Reynolds number, more sophisticated methods for generating good initial guesses are often needed. One such class of methods are *continuation methods*. The basic idea is to use information gained at a value of the Reynolds number for which it is "easy" to solve the discrete equations to help generate a good initial guess for the solution of the discrete equations at a "hard" value of the Reynolds number. In general, continuation methods are made up of a sequence of predictor steps, each of which is followed by a corrector step. The latter is usually chosen to be some iterative method such as Newton's method or a composite simple iteration-Newton's method. A predictor step may be viewed as a method for generating an initial guess for the corrector step that follows.

We begin by briefly describing continuation methods in the context of general mappings from $\mathbb{R}^{s+1} \to \mathbb{R}^s$. For detailed discussions of continuation methods in this general framework, see den Heijer and Rheinboldt [1981], Keller [1978 and 1987], and Rheinboldt [1980]; also, see Ortega and Rheinboldt [1970].

One is given the mapping $\mathbf{G}(\mathbf{x}, \lambda)$: $\mathbb{R}^{s+1} \to \mathbb{R}^s$ where $\mathbf{x} \in \mathbb{R}^s$ and $\lambda \in \mathbb{R}$. Here λ may be a naturally appearing parameter, e.g., the Reynolds number for the discrete Navier–Stokes equations, or may be an artificially introduced parameter such as the arc length along a solution branch, or it may even be one of the components of the desired solution vector. One wishes to find $\mathbf{x}^*$ such that $\mathbf{G}(\mathbf{x}^*, \lambda^*) = \mathbf{0}$ for some given value $\lambda = \lambda^*$. It is assumed that a solution $\tilde{\mathbf{x}}$ satisfying $\mathbf{G}(\tilde{\mathbf{x}}, \tilde{\lambda}) = \mathbf{0}$ is known, where without loss of generality, one may assume that $\tilde{\lambda} < \lambda^*$. Starting with the known solution $\tilde{\mathbf{x}}$ at $\lambda = \tilde{\lambda}$, the solution at $\lambda = \lambda^*$ is obtained by following the solution curve $\mathbf{x}(\lambda)$ for $\lambda \in [\tilde{\lambda}, \lambda^*]$ where $\mathbf{G}(\mathbf{x}(\lambda), \lambda) = \mathbf{0}$. This curve is determined through a predictor-corrector procedure. Suppose $\mathbf{x}(\lambda)$ is known for some value $\lambda \in [\tilde{\lambda}, \lambda^*)$; since $\mathbf{x}(\tilde{\lambda}) = \tilde{\mathbf{x}}$ is presumed known, this supposition is not vacuous. One first predicts an approximation $\mathbf{x}_0$ for $\mathbf{x}(\lambda_+)$ where $\lambda_+ > \lambda$. The most common choices for the predictor are either to use the old value of $\mathbf{x}$, i.e.,

$$\mathbf{x}_0 = \mathbf{x}(\lambda) \tag{8.1}$$

or to follow the tangent line to the curve at the known point $[\lambda, \mathbf{x}(\lambda)]$, i.e.,

$$\mathbf{x}_0 = \mathbf{x}(\lambda) + \Delta\lambda \frac{d\mathbf{x}}{d\lambda}(\lambda) \quad \text{where } \Delta\lambda = \lambda_+ - \lambda. \tag{8.2}$$

In (8.2) the tangent vector $d\mathbf{x}/d\lambda$ is obtained by differentiating the defining relation $\mathbf{G}(\mathbf{x}(\lambda), \lambda) = \mathbf{0}$, i.e.,

$$\frac{d\mathbf{x}}{d\lambda}(\lambda) = -\{\mathbf{G}_\mathbf{x}(\mathbf{x}(\lambda), \lambda)\}^{-1}\mathbf{G}_\lambda(\mathbf{x}(\lambda), \lambda), \tag{8.3}$$

where $\mathbf{G}_\mathbf{x}(\mathbf{x}, \cdot)$ denotes the $s \times s$ Jacobian matrix of $\mathbf{G}$ with respect to $\mathbf{x}$ and $\mathbf{G}_\lambda(\cdot, \lambda)$ denotes the s- vector whose components are the partial derivatives, with respect to λ, of the corresponding components of $\mathbf{G}$. One easily observes that (8.2) is simply a forward Euler approximation to the solution of the ordinary differential equation (8.3). Of course, in order for (8.2) to be of use, one must have that $\mathbf{G_x}$ is invertible for $\lambda \in [\tilde{\lambda}, \lambda^*]$. If this is not the case, e.g., whenever one is at or near a turning point or bifurcation point of the solution curve, one must invoke special procedures, e.g., introduce an artificial parameter or switch parameters. It is assumed that the solution curve $\mathbf{x}(\lambda)$ for $\lambda \in [\tilde{\lambda}, \lambda^*]$ is a section of a nonsingular branch, i.e., for $\lambda \in [\tilde{\lambda}, \lambda^*]$ $\mathbf{G_x}(\mathbf{x}(\lambda), \lambda)$ is invertible and, moreover, $\{\mathbf{G_x}(\mathbf{x}(\lambda), \lambda)\}^{-1}$ is uniformly bounded with respect to λ.

Once $\mathbf{x}_0$ is obtained through, e.g., (8.1) or (8.2), in principle an arbitrarily close approximation to the exact solution $\mathbf{x}(\lambda_+)$ is obtained through some iterative procedure, e.g., Newton's method, using $\mathbf{x}_0$ for the initial guess. If this corrective step is to produce a convergent sequence of approximations to $\mathbf{x}(\lambda_+)$, $\mathbf{x}_0$ must, of course, be within the attraction region at $\lambda = \lambda_+$ for the chosen iterative scheme. Once $\mathbf{x}(\lambda_+)$ has been determined, the procedure may be repeated, starting with this value to obtain $\mathbf{x}$ at an even larger value of λ.

In terms of the discrete Navier–Stokes equations, a single step of the above continuation algorithm is given as follows. Having determined a solution $\mathbf{u}^h(Re) \in \mathbf{V}_0^h$ and $p^h(Re) \in S_0^h$ at a particular value Re of the Reynolds number, an initial guess $\mathbf{u}_0$ for the corrective iteration at $Re_+ > Re_0$ is defined by either

$$\mathbf{u}_0 = \mathbf{u}^h = \mathbf{u}^h(Re) \tag{8.4}$$

or

$$\mathbf{u}_0 = \mathbf{u}^h + \mathbf{u}^{(1)} \Delta Re, \tag{8.5}$$

where $\Delta Re = Re_+ - Re$ and $\mathbf{u}^{(1)}(Re) = d\mathbf{u}^h/dRe \in \mathbf{V}_0^h$ is the solution of

$$a(\mathbf{u}^{(1)}, \mathbf{v}) + c(\mathbf{u}^{(1)}, \mathbf{u}^h, \mathbf{v}) + c(\mathbf{u}^h, \mathbf{u}^{(1)}, \mathbf{v}) + b(\mathbf{v}, p^{(1)})$$

$$= \frac{1}{Re} a(\mathbf{u}^h, \mathbf{v}) \quad \text{for all } \mathbf{v} \in \mathbf{V}_0^h \tag{8.6}$$

and

$$b(\mathbf{u}^{(1)}, q) = 0 \quad \text{for } q \in S_0^h, \tag{8.7}$$

where we assume $\nu = 1/Re$, i.e., the variables have been nondimensionalized. Note that the left-hand side of (8.6)–(8.7) is exactly that of (6.3)–(6.4) when $\mathbf{u}^{(m-1)} \approx \mathbf{u}^{(m)} \approx \mathbf{u}^h(Re)$, i.e., the linear system used to find $\mathbf{u}^{(1)}(Re)$ has the same coefficient matrix as the linear system encountered in the last step of Newton's method for computing $\mathbf{u}^h(Re)$. Thus, if the corrective iteration is chosen to be Newton's method, $\mathbf{u}^{(1)}(Re)$ may be computed through a forward and backward solve procedure using the (factored) coefficient matrix that determined the last Newton iterate at that value of the Reynolds number.

Once a predictor method, e.g., (8.1) or (8.2), and a corrective iterative method, e.g., Newton's method, are chosen, to complete the description of a single continuation step, a method for choosing the step size ΔRe must be defined. In practice, it is found that whenever Re is not near a value of the Reynolds number at which a bifurcation or turning point occurs, the step size in the Reynolds number ΔRe may be rather large, e.g., for $Re \gg 1$ roughly proportional to the Reynolds number Re itself. *This is independent of which predictor algorithm is used* (Gunzburger and Peterson [1989]). Thus in these situations, it does not seem to pay to use the more complicated predictor (8.2).

On the other hand, near bifurcation and turning points the allowable step size ΔRe will become much smaller and it may be advantageous to use the predictor (8.2). Moreover, it is difficult, if not impossible, to pass through such points using algorithms that involve continuation in the Reynolds number. In order to avoid these difficulties it is recommended that near such points one switch to some other continuation parameter such as the arc length along the solution curve or some judiciously chosen component of the solution. See the cited references for details.

8.2. The Reduced Basis Method

The discretization algorithms discussed in Chapters 1–5 make use of basis functions that have little to do with the particular problem whose solutions we are trying to approximate in the sense that the same basis functions are used in finding approximate solutions to other very different problems. Thus the same piecewise polynomial spaces are used to find approximate solutions of the Navier–Stokes equations as are used for second-order elliptic equations, hyperbolic equations, etc. The analogous observation can be made about other classes of discretization techniques. For example, for spectral methods one uses Chebyshev polynomials for all sorts of different problems. The reduced basis method, which can be defined in the context of finite element, finite difference, spectral, etc. methods, attempts to use basis functions that have something to do with the problem one is trying to solve. The hope is that by using such basis functions one can get away with using less of them. A discussion of the use of this method for approximating viscous incompressible flows can be found in Peterson [1989]. The method has been previously developed and applied to structural mechanics problems; see Noor [1980] and the references cited there and in Peterson [1989].

First, we describe how to generate the set of "reduced" basis vectors; then we will show how they are used, and finally, we will remark on the reasons why the method can be effective.

There are at least two different mechanisms for generating the reduced basis vectors. Consider the discrete system (2.1)–(2.2), which we rewrite in the form

$$\frac{1}{Re} \tilde{a}(\mathbf{u}^h, \mathbf{v}^h) + c(\mathbf{u}^h, \mathbf{u}^h, \mathbf{v}^h) + b(\mathbf{v}^h, p^h) = (\mathbf{f}, \mathbf{v}^h) \quad \text{for all } \mathbf{v}^h \in V_0^h$$

(8.8)

and

$$b(\mathbf{u}^h, q^h) = 0 \quad \text{for all } q^h \in S_0^h \tag{8.9}$$

in order to highlight the only explicit appearance of the parameter Re. Of course, $\tilde{a}(\cdot, \cdot) = Re \, a(\cdot, \cdot)$ and we are setting $Re = 1/\nu$, i.e., the variables are appropriately nondimensionalized. We will denote the solution of the discrete equations (8.8)–(8.9) for a given value Re of the Reynolds number by $\mathbf{u}^h(Re)$ and $p^h(Re)$.

One set of reduced basis vectors is determined by solving (8.8)–(8.9) for different values of the Reynolds number. Thus, given a set of values for the Reynolds number $\{Re_m : m = 0, ..., M\}$, we solve (8.8)–(8.9) $M + 1$ times to determine the set $\{\hat{\mathbf{u}}_m, \hat{p}_m : m = 0, ..., M\}$, where $\hat{\mathbf{u}}_m = \mathbf{u}^h(Re_m)$ and $\hat{p}_m = p^h(Re_m)$ for $m = 0, ..., M$. We then set $V_L^M = \text{span}\{\hat{\mathbf{u}}_m : m = 0, ..., M\} \subset V_0^h$ and $S_L^M = \text{span}\{\hat{p}_m : m = 0, ..., M\} \subset S_0^h$, where the inclusions follow, since clearly $\hat{\mathbf{u}}_m \in V_0^h$ and $\hat{p}_m \in S_0^h$ for all m. For obvious reasons, the basis functions $(\hat{\mathbf{u}}_m, \hat{p}_m)$ generated in this manner are called *Lagrange reduced basis functions*.

A second way to generate reduced basis functions is to first solve, at a given value Re_0 of the Reynolds number, (8.8)–(8.9) for $\mathbf{u}_0 = \mathbf{u}^h(Re_0)$ and $p_0 = p^h(Re_0)$. Then, for $m = 1, ..., M$, one differentiates (8.8)–(8.9) m times with respect to Re and then evaluates the results at $Re = Re_0$.

Setting

$$\mathbf{u}_m = \frac{d^m \mathbf{u}^h}{d(Re)^m}(Re_0) \quad \text{and} \quad p_m = \frac{d^m p^h}{d(Re)^m}(Re_0) \quad \text{for } m = 0, ..., M,$$

we then have, for $m = 1, ..., M$, that $\mathbf{u}_m \in \mathbf{V}_0^h$ and $p_m \in S_0^h$ satisfy

$$\frac{1}{Re_0}\, \tilde{a}(\mathbf{u}_m, \mathbf{v}^h) + c(\mathbf{u}_m, \mathbf{u}_0, \mathbf{v}^h) + c(\mathbf{u}_0, \mathbf{u}_m, \mathbf{v}^h) + b(\mathbf{v}^h, p_m)$$

$$= G_m[\mathbf{u}_0, \mathbf{u}_1, ..., \mathbf{u}_{m-1}; Re_0; \mathbf{v}^h) \quad \text{for all } \mathbf{v}^h \in \mathbf{V}_0^h \quad (8.10)$$

and

$$b(\mathbf{u}_m, q^h) = 0 \quad \text{for all } q^h \in S_0^h, \quad (8.11)$$

where $G_m(\cdot; \cdot; \cdot)$ is "easily" determined through the differentiation process. For example,

$$G_1(\mathbf{u}_0; Re_0; \mathbf{v}^h) = \frac{1}{Re_0^2}\, \tilde{a}(\mathbf{u}_0, \mathbf{v}^h)$$

and

$$G_2(\mathbf{u}_0, \mathbf{u}_1; Re_0; \mathbf{v}^h) = \frac{2}{Re_0^2}\, \tilde{a}(\mathbf{u}_1, \mathbf{v}^h) - \frac{2}{Re_0^3}\, \tilde{a}(\mathbf{u}_0, \mathbf{v}^h) - 2c(\mathbf{u}_1, \mathbf{u}_1, \mathbf{v}^h).$$

Note that (8.10)–(8.11) is linear in $(\mathbf{u}_m, p_m)$; in the usual manner, they define a sequence of linear systems of equations for $(\mathbf{u}_m, p_m)$, $m = 1, ..., M$. All of these linear systems have the same left-hand side so that, e.g., they all may be solved with a single matrix factorization. In fact, if $\mathbf{u}_0 = \mathbf{u}^h(Re_0)$ is found using Newton's method, then the single matrix that needs to be factored to determine all $(\mathbf{u}_m, p_m)$ for $m \geq 1$ can be taken to be the matrix that determined the last Newton iterate; in this case, $(\mathbf{u}_m, p_m)$ for $m \geq 1$ can be determined with no additional factorizations over those needed to determine $\mathbf{u}_0$, i.e., to solve (8.8)–(8.9). Clearly $\mathbf{u}_0 = \hat{\mathbf{u}}_0$ and $p_0 = \hat{p}_0$, but, at least in the context of direct matrix solutions, $(\mathbf{u}_m, p_m)$ are less expensive to generate than are $(\hat{\mathbf{u}}_m, \hat{p}_m)$.

We set

$$\mathbf{V}_T^M = \text{span}\{\mathbf{u}_m : m = 0, ..., M\} \subset \mathbf{V}_0^h$$

and

$$S_T^M = \text{span}\{p_m : m = 0, ..., M\} \subset S_0^h.$$

Again, for obvious reasons, the basis functions $(\mathbf{u}_m, p_m)$ generated in this manner are called *Taylor reduced basis functions*.

There are other ways to generate reduced basis functions. For example, one can use least square approximations in a manner analogous to the above use of Lagrange interpolants and Taylor polynomials. The most popular choice in structural dynamics applications has been the Taylor basis, and that is the one on which we will focus our attention.

Now that we have shown how to generate reduced basis functions, or specifically, the subspaces $\mathbf{V}_T^M$ and S_T^M, we now proceed to show how these are used to determine some kind of approximation to $\mathbf{u}$ and p at some other value of the Reynolds number $Re^* \geq Re_0$. What is *not* done is to compute the Taylor polynomials

$$\mathbf{u}_T^M(Re^*) = \sum_{m=0}^{M} \frac{(Re^* - Re_0)^m}{m!} \mathbf{u}_m \quad \text{and} \quad p_T^M(Re^*) = \sum_{m=0}^{M} \frac{(Re^* - Re_0)^m}{m!} p_m.$$

$$(8.12)$$

Instead, we seek $\mathbf{u}_R^M(Re^*) \in \mathbf{V}_T^M = \mathrm{span}\{\mathbf{u}_m : m = 0, ..., M\} \subset \mathbf{V}_0^h$ and $p_R^M(Re^*) \in S_T^M = \mathrm{span}\{p_m : m = 0, ..., M\} \subset S_0^h$ such that

$$\frac{1}{Re^*} \tilde{a}(\mathbf{u}_R^M, \mathbf{v}_R^M) + c(\mathbf{u}_R^M, \mathbf{u}_R^M, \mathbf{v}_R^M) + b(\mathbf{v}_R^M, p_R^M) = (\mathbf{f}, \mathbf{v}_R^M)$$

$$\text{for all } \mathbf{v}_R^M \in \mathbf{V}_T^M \qquad (8.13)$$

and

$$b(\mathbf{u}_R^M, q_R^M) = 0 \quad \text{for all } q_R^M \in S_T^M, \qquad (8.14)$$

i.e., we discretize the Navier–Stokes equations using the reduced spaces $\mathbf{V}_T^M$ and S_T^M.

Actually, the reduced basis problem is simpler than (8.13)–(8.14). Note that due to (8.9) and (8.11), we have $\mathbf{u}_m \in \mathbf{Z}^h$ so that $\mathbf{V}_T^M \subset \mathbf{Z}^h$. Thus (8.14) is automatically satisfied and (8.13) reduces to

$$\frac{1}{Re^*} \tilde{a}(\mathbf{u}_R^M, \mathbf{v}_R^M) + c(\mathbf{u}_R^M, \mathbf{u}_R^M, \mathbf{v}_R^M) = (\mathbf{f}, \mathbf{v}_R^M) \quad \text{for all } \mathbf{v}_R^M \in \mathbf{V}_T^M.$$

$$(8.15)$$

Thus the computation of the reduced basis approximation of the velocity uncouples from that of the pressure. Note that the fact that $\mathbf{V}_T^M \subset \mathbf{Z}^h$ also relieves us of any worry about satisfying the div-stability condition over the reduced subspaces.

It is clear, due to the fact that the reduced basis functions $\mathbf{u}_m$ are globally supported, that (8.15) is equivalent to a *dense* nonlinear system of equations. This should be contrasted with (8.8)–(8.9) which, due to the small support of the usual finite element basis functions, is a *sparse* nonlinear system. Thus (8.15) is less costly to solve only if the dimension of the reduced basis spaces V_T^M is small, i.e., if M is small, relative to the dimensions of V_0^h and S_0^h.

Let us assume that the target Reynolds number Re^* is large compared to the Reynolds number Re_0 at which we computed the reduced basis vectors. There are two different uses the reduced basis solution $u_R^M(Re^*)$ can be put to. One may simply use it as the sought for approximation at $Re = Re^*$, or one may use it as an initial guess for some iterative method for solving (8.8)–(8.9) for $u^h(Re^*)$ and $p^h(Re^*)$. Thus, in the context of continuation methods, we have a total of three choices of how to determine an approximation of $u(Re^*)$ and $p(Re^*)$, the exact solution of the Navier–Stokes equations at the target Reynolds number.

First, starting with the solution $u^h(Re_0)$ and $p^h(Re_0)$ of (8.8)–(8.9) at Re_0, we can use a continuation method for (8.8)–(8.9) to find a solution of this system at Re^*. This means that for each of the intermediate Reynolds numbers between Re_0 and Re^*, as well as at Re_0 and Re^*, one must solve the nonlinear system (8.8)–(8.9) whose size is the sum of the dimensions of V_0^h and S_0^h.

Instead, starting with $u^h(Re_0) \in V_T^M \subset V_0^h$ one could use a continuation method for (8.15) to find a solution of this system at Re^* and then use that solution as an initial guess for some iterative method for (8.8)–(8.9). This means that we solve the system (8.8)–(8.9) only for Re_0 and Re^*, while at the intermediate Reynolds numbers we solve the system (8.15). If the initial guess $u_R^M(Re^*)$ supplied by the reduced basis method is sufficiently good so that the iterative scheme for solving (8.8)–(8.9) with $Re = Re^*$ converges, then the solution obtained in this second case is the same as that obtained in the first case.

In the third case, we just take the solution of (8.15) at Re^* to be the sought for approximation so that we only solve (8.8)–(8.9) at Re_0. It should also be pointed out that in the structural mechanics community, wherein the reduced basis method has undergone substantial algorithmic development, it is viewed mostly as a method for generating approximate solutions, and not to generate initial guesses for some other method. See Noor [1980] and the references cited in Peterson [1989].

Now one may ask, since the solutions obtained in the first two cases are the same, which one is less costly to produce? One may also ask how good is the solution obtained in the third case, and how costly is it to produce? To repeat an observation made above, the cost effectiveness of the second and third techniques, relative to the first, depends on being able to make them work with a "small" choice of M. Making the methods work can mean slightly different things. For the second method, we only require the reduced basis solution at Re^* to be good enough so that an iterative scheme for (8.8)–(8.9) converges. For the third method, that solution itself has to be good enough to be a meaningful approximation to the exact solution.

Of course, one cannot continue using the reduced basis vectors produced at Re_0 to compute a solution in the reduced basis space at arbitrarily high values of the Reynolds; from time to time, the reduced basis vectors have to be recomputed and subsequent solutions are sought in the new reduced basis space.

Computational evidence in both structural and fluid mechanics applications indicate that M may be chosen *very small*, i.e., somewhere in the range of 5 to 15. Furthermore, even for Reynolds numbers considerably larger than the Reynolds number Re_0 at which the reduced basis vectors are determined, the reduced basis solution itself often turns out to be a good approximation. At this point there is no adequate rigorous explanation of why these observations are true. However, there are some indications, based on both computations and analyses, of what is behind the apparent success of the method.

It seems that the main reason why M may be chosen so small is that the reduced basis vectors quickly become nearly linearly dependent. Thus the spaces $\mathbf{V}_T^M = \mathrm{span}\{\mathbf{u}_0, ..., \mathbf{u}_M\}$ and $\mathbf{V}_T^{M+k} = \mathrm{span}\{\mathbf{u}_0, ..., \mathbf{u}_{M+k}\}$, $k > 0$, are nearly the same in the sense that adding the extra vectors $\mathbf{u}_{M+1}, ..., \mathbf{u}_{M+k}$ does not appreciably improve the ability to approximate. This does not help in keeping the number of terms small in a Taylor series approximation such as (8.12), since there the coefficients are predetermined functions of Re^*. However, the solution of (8.15) is in some sense a best approximation, or at least its error can be bounded by the error in a best approximation. Furthermore, due to the near linear dependence of the vectors $\mathbf{u}_{M+1}, ..., \mathbf{u}_{M+k}$ with respect to the vectors belonging to $\mathbf{V}_T^M$, best approximations out of $\mathbf{V}_T^M$ are very nearly the same as those out of $\mathbf{V}_T^{M+k}$.

Another observation concerning the set of reduced basis vectors is that their *size* decreases very rapidly as m increases. In fact, there is evidence (see Gunzburger and Peterson [1989] and Peterson [1989]) that $|\mathbf{u}_m|_1 = O(Re_0^{-m}|\mathbf{u}_0|_1)$, at least away from bifurcation and turning points. Thus, even for moderate values of Re_0, we have a rapid decrease in the size of the reduced basis vectors. What this means is that even the Taylor polynomial (8.12) provides a good approximation for large steps in the Reynolds number. The solution of (8.15), being a better approximation than the Taylor polynomial, remains a good approximation for even larger values of Re^*.

III

TIME-DEPENDENT PROBLEMS

We now turn our attention towards time-dependent problems and their approximation by finite element methods. Fortunately, just about everything that we have said about stationary problems is relevant to the time-dependent case. The main reason for this is that the most effective algorithms for treating time-dependent problems can be defined through a process wherein the spatial and temporal discretizations are separated. As we shall see below, spatial discretizations are effected by finite element methods, while temporal discretizations are more easily brought about by finite difference methods.

9

A Weak Formulation and Spatial Discretizations

9.1. A Weak Formulation for the Time-Dependent Problem

In this chapter we consider finite element methods for the time-dependent Navier–Stokes equations

$$\frac{\partial \mathbf{u}}{\partial t} + \mathbf{u} \cdot \operatorname{grad} \mathbf{u} + \operatorname{grad} p - \nu \Delta \mathbf{u} = \mathbf{f} \quad \text{in } \Omega \times (0, T], \tag{9.1}$$

$$\operatorname{div} \mathbf{u} = 0 \quad \text{in } \Omega \times (0, T], \tag{9.2}$$

$$\mathbf{u} = \mathbf{0} \quad \text{on } \Gamma \times (0, T], \tag{9.3}$$

and

$$\mathbf{u}(0, \mathbf{x}) = \mathbf{u}_0(\mathbf{x}) \quad \text{in } \Omega, \tag{9.4}$$

where $\mathbf{u}(t, \mathbf{x})$ and $p(t, \mathbf{x})$ denote the velocity and pressure, respectively, $\mathbf{f}(t, \mathbf{x})$ the body force per unit mass, ν the kinematic viscosity (or, should $\mathbf{u}$ and p be nondimensionalized, the inverse of the Reynolds number), and Ω a bounded open set of $\mathbb{R}^2$ or $\mathbb{R}^3$. Furthermore, T is a given positive constant and $\mathbf{u}_0(\mathbf{x})$ is a given function. Note that due to the absence of any time derivative of the pressure, no initial condition for

117

the pressure is needed. Discretizations of problems involving other boundary conditions and/or other formulations of the viscous term may be defined by making changes to the algorithms discussed in this chapter similar to those for the stationary case discussed in Chapter 4.

The discretization with respect to the spatial variables is effected by a finite element method. As was the case for the stationary problem, the main difficulty is due to the incompressibility constraint (9.2). Fortunately, our discussion in Chapters 2 and 3 concerning div-stability is relevant here, and we will be able to use any of the div-stable elements discussed in those chapters. Due to the cylindrical nature of the domain $\Omega \times [0, T]$, the discretization with respect to the time variable is most easily effected via *finite difference methods*, and it is such discretizations that will be considered here.

As usual, we begin by considering a weak form of the problem (9.1)–(9.4). To this end, we need to define some additional function spaces. For each t, we view $\mathbf{u}$ and p as functions of time that take on values in a function space. For example, if X is a Banach space with norm $\|\cdot\|_X$, we denote by $L^r(0, T; X)$ the space of strongly measurable maps $\phi \colon [0, T] \to X$ such that

$$\|\phi\|_{L^r(0,T;X)} = \left(\int_0^T \|\phi\|_X^r \, dt \right)^{1/r} < \infty \qquad \text{if } 1 \le r < \infty$$

or

$$\|\phi\|_{L^\infty(0,T;X)} = \operatorname*{ess\,sup}_{0 < t < T} \|\phi\|_X < \infty \qquad \text{if } r = \infty,$$

e.g., the $L^r(0, T)$-norm of the spacial X-norm of ϕ is bounded. We will need to use the space

$$\mathbf{H} = \{\mathbf{u} \in \mathbf{L}^2(\Omega) : \operatorname{div} \mathbf{u} = 0 \text{ in } \Omega; \quad \mathbf{u} = 0 \text{ on } \Gamma\},$$

where the constraints on the functions belonging to $\mathbf{H}$ should be properly interpreted, e.g., $\operatorname{div} \mathbf{u} = 0$ holds in $H^{-1}(\Omega)$.

We use the forms $a(\cdot, \cdot)$, $b(\cdot, \cdot)$, and $c(\cdot, \cdot, \cdot)$ defined by (1.6), (1.7), and (1.8), respectively, and we let $(\cdot, \cdot)$ denote the inner product in $\mathbf{L}^2(\Omega)$.

Having defined the above spaces and forms, given

$$\mathbf{f} \in L^2(0, T; \mathbf{H}^{-1}(\Omega)) \qquad \text{and} \qquad \mathbf{u}_0 \in \mathbf{H},$$

we seek

$$\mathbf{u} \in L^2(0, T; \mathbf{H}_0^1(\Omega)) \cap L^\infty(0, T; \mathbf{H})$$

and
$$p \in L^2[0, T; L_0^2(\Omega)]$$
such that

$$\left(\frac{\partial \mathbf{u}}{\partial t}, \mathbf{v}\right) + a(\mathbf{u}, \mathbf{v}) + c(\mathbf{u}, \mathbf{u}, \mathbf{v}) + b(\mathbf{v}, p) = (\mathbf{f}, \mathbf{v}) \qquad \text{for all } \mathbf{v} \in \mathbf{H}_0^1(\Omega),$$
$$(9.5)$$

$$b(\mathbf{u}, q) = 0 \qquad \text{for } q \in L_0^2(\Omega), \qquad (9.6)$$

and

$$\mathbf{u}(0, \mathbf{x}) = \mathbf{u}_0(\mathbf{x}) \qquad \text{for } \mathbf{x} \in \Omega, \qquad (9.7)$$

where (9.5) and (9.6) hold, on $(0, T)$, in the sense of distributions. Note that the initial condition is required to satisfy the boundary condition and the incompressibility constraint.

For details concerning (9.5)–(9.7), as well as for results concerning the existence, uniqueness, and regularity of solutions of these equations, consult Girault and Raviart [1979], Ladyzhenskaya [1969], and Temam [1979 and 1983]. We do make note of the "existence-uniqueness gap" in three dimensions, namely, that one can prove that solutions are unique only within a class of functions that are smoother than the class in which one can prove that solutions exist.

9.2. Spatial Discretizations

The discretization with respect to the spatial variables is effected exactly as described in Chapter 2 for the stationary case. We choose subspaces $\mathbf{V}_0^h \subset \mathbf{H}_0^1(\Omega)$ and $S_0^h \subset L_0^2(\Omega)$ and seek a $\mathbf{u}^h(t, \cdot) \in \mathbf{V}_0^h$ and $p^h(t, \cdot) \in S_0^h$ such that

$$\left(\frac{\partial \mathbf{u}^h}{\partial t}, \mathbf{v}^h\right) + a(\mathbf{u}^h, \mathbf{v}^h) + c(\mathbf{u}^h, \mathbf{u}^h, \mathbf{v}^h) + b(\mathbf{v}^h, p^h)$$

$$= (\mathbf{f}, \mathbf{v}^h) \qquad \text{for all } \mathbf{v}^h \in \mathbf{V}_0^h \text{ and for } 0 < t \le T, \qquad (9.8)$$

$$b(\mathbf{u}^h, q^h) = 0 \qquad \text{for all } q^h \in S_0^h \text{ and for } 0 < t \le T, \qquad (9.9)$$

and

$$\mathbf{u}^h(0, \mathbf{x}) = \mathbf{u}_0^h \in \mathbf{V}_0^h \text{ for } \mathbf{x} \in \Omega, \qquad (9.10)$$

where $\mathbf{u}_0^h$ is an approximation to the initial function $\mathbf{u}_0(\mathbf{x})$.

After choosing bases for $\mathbf{V}_0^h$ and S_0^h, (9.8)–(9.10) are equivalent to a system of nonlinear ordinary differential equations with linear algebraic constraints. Indeed, if

$$p^h(t, \mathbf{x}) = \sum_{j=1}^{J} \alpha_j(t) q_j(\mathbf{x}) \quad \text{and} \quad \mathbf{u}^h(t, \mathbf{x}) = \sum_{k=1}^{K} \beta_k(t) \mathbf{v}_k(\mathbf{x}),$$

where $\{\mathbf{v}_k(\mathbf{x})\}$, $k = 1, ..., K$, and $\{q_j(\mathbf{x})\}$, $j = 1, ..., J$, denote bases for $\mathbf{V}_0^h$ and S_0^h, respectively, then (9.8)–(9.10) are equivalent to the system of ordinary differential equations

$$\sum_{k=1}^{K} (\mathbf{v}_k, \mathbf{v}_\ell) \frac{d\beta_k}{dt} + \sum_{k=1}^{K} a(\mathbf{v}_k, \mathbf{v}_\ell) \beta_k(t) + \sum_{k,m=1}^{K} c(\mathbf{v}_m, \mathbf{v}_k, \mathbf{v}_\ell) \beta_k(t) \beta_m(t)$$

$$+ \sum_{j=1}^{J} b(\mathbf{v}_\ell, q_j) \alpha_j(t) = (\mathbf{f}, \mathbf{v}_\ell) \quad \text{for } \ell = 1, ..., K, \tag{9.11}$$

with initial data $\beta_k(0)$, $k = 1, ..., K$, which satisfy

$$\sum_{k=1}^{K} \mathbf{v}_k \beta_k(0) = \mathbf{u}_0^h \tag{9.12}$$

and are subject to the linear algebraic constraints

$$\sum_{k=1}^{K} b(\mathbf{v}_k, q_i) \beta_k(t) = 0 \quad \text{for } i = 1, ..., J. \tag{9.13}$$

The system of ordinary differential equations (9.11), or equivalently (9.8), may be discretized by any of the numerous known ordinary differential equation solvers. We will focus, in the next chapter, on a few such methods that we feel are the most useful in the context of the Navier–Stokes equations.

When discussing discretizations with respect to time, it will be convenient to rewrite the semi-discrete system (9.11) as

$$\left(\frac{\partial \mathbf{u}^h}{\partial t}, \mathbf{v}^h \right) = \mathbf{F}(\mathbf{f}, \mathbf{u}^h, p^h; \mathbf{v}^h) \quad \text{for all } \mathbf{v}^h \in \mathbf{V}_0^h, \tag{9.14}$$

where the linear functional $\mathbf{F}(\cdot, \cdot, \cdot; \mathbf{v}^h)$ is defined, for any $\mathbf{u}^h \in \mathbf{V}_0^h$ and $p^h \in S_0^h$ and any $\mathbf{f}$, by

$$\mathbf{F}(\mathbf{f}, \mathbf{u}^h, p^h; \mathbf{v}^h) = (\mathbf{f}, \mathbf{v}^h) - a(\mathbf{u}^h, \mathbf{v}^h) - c(\mathbf{u}^h, \mathbf{u}^h, \mathbf{v}^h)$$

$$- b(\mathbf{v}^h, p^h) \quad \text{for all } \mathbf{v}^h \in \mathbf{V}_0^h. \tag{9.15}$$

For all the methods discussed in Chapter 10, the spatial errors in the velocity approximation are the expected ones, i.e., exactly what one would obtain in an analogous stationary calculation using the same finite element spaces. The only requirement is that the starting value $\mathbf{u}^0$ be a sufficiently close approximation to $\mathbf{u}_0$, i.e., no worse than the spatial accuracy of the analogous stationary method. In fact, for most of the methods it is not actually required that $\mathbf{u}^0 \in \mathbf{V}_0^h$ so that one may, in these cases, choose $\mathbf{u}^0 = \mathbf{u}_0$ where $\mathbf{u}_0$ denotes the given initial data (9.4).

10

Time Discretizations

We will consider time discretization algorithms that fall into one of four classes of methods, namely single-step and multistep methods of both fully implicit and semi-implicit type. Multistep methods are more cumbersome to implement than are single-step methods, but the former yield higher time accuracy. Fully implicit methods are likewise more cumbersome to implement than are semi-implicit methods, but the former have better stability properties, i.e., fully implicit methods are usually unconditionally stable while the semi-implicit methods we will discuss are only conditionally stable.

Note that the time derivative $d\beta_k/dt$ in the ℓth equation of (9.11) is multiplied by the element $(\mathbf{v}_k, \mathbf{v}_\ell)$ of the Gram matrix of the basis $\{\mathbf{v}_k\}$ so that normally explicit methods, e.g., Euler's method, still require the solution of a matrix problem at every time step. However, the symmetric, positive definite coefficient matrix does not change from time step to time step, so that it needs to be factored only once. On the other hand, explicit methods are not in common use for time discretizations of the Navier–Stokes equations because of their severe stability restriction.

The time interval $[0, T]$ is subdivided into M intervals of uniform length δ so that $\delta = T/M$. Throughout, $\mathbf{u}^m$ and p^m, $m = 0, ..., M$, will respectively denote approximations to $\mathbf{u}^h(m\delta, \mathbf{x})$ and $p^h(m\delta, \mathbf{x})$ where $\mathbf{u}^h$ and p^h denote the solution of (9.8)–(9.9). Likewise, for $m = 0, \ldots, M$, $k = 1, \ldots, K$, and $j = 1, \ldots, J$, α_j^m and β_k^m denote approximations to $\alpha_j(m\delta)$ and $\beta_k(m\delta)$, respectively, where α_j, $j = 1, \ldots, J$, and β_k, $k = 1, \ldots, K$, denote the solution of (9.11) and (9.13). Also, throughout, $\mathbf{f}^m = \mathbf{f}(m\delta, \mathbf{x})$.

10.1. Single-Step Fully Implicit Methods

Perhaps the simplest choice of discretization for (9.14) is the "θ-method." Given $\mathbf{u}^0$, which may be chosen to be $\mathbf{u}_0$, $\{\mathbf{u}^m, p^m\}$ for $m = 1, \ldots, M$ are determined from

$$\frac{1}{\delta}(\mathbf{u}^m - \mathbf{u}^{m-1}, \mathbf{v}^h) = \mathbf{F}(\mathbf{f}_\theta^m, \mathbf{u}_\theta^m, p_\theta^m; \mathbf{v}^h) \quad \text{for all } \mathbf{v}^h \in V_0^h \quad (10.1)$$

and

$$b(\mathbf{u}^m, q^h) = 0 \quad \text{for all } q^h \in S_0^h, \quad\quad (10.2)$$

where

$$\mathbf{u}_\theta^m = \theta\mathbf{u}^m + (1 - \theta)\mathbf{u}^{m-1} \quad \text{and} \quad p_\theta^m = \theta p^m + (1 - \theta)p^{m-1}, \quad (10.3)$$

and likewise for $\mathbf{f}_\theta^m$. The scheme (10.1) is the explicit Euler method when $\theta = 0$; the implicit Euler, or backward differentiation, method when $\theta = 1$; and the Crank–Nicholson, or trapezoidal rule, method for $\theta = 1/2$. The latter two methods are the ones of greatest interest.

It seems that the scheme (10.1)–(10.2) requires an initial condition for the pressure whenever $\theta \neq 1$, since p^1 is obtained from p_θ^0 and p^0 through (10.3). However, if one is only interested in the velocity field, one never need bother with computing p^1, since the latter will not be used in the velocity computation at the next time level, i.e., $t = 2\delta$. Furthermore, for $\theta \neq 1/2$ this scheme is only first-order accurate with respect to δ so that, since by Taylor's theorem, $p^1 = p_\theta^0 + O[(1 - \theta)\delta] + O(\delta^2)$, one may take p_θ^0 as the pressure approximation at $t = \delta$ and again never need p^0. For $\theta = 1/2$ this procedure results in a loss of accuracy in the pressure; this is not so bad since the spatial errors for the pressure are usually larger than those for the velocity anyway.

Clearly (10.1)–(10.2) is a system of nonlinear algebraic equations. In order to minimize the cost of computing each pair $(\mathbf{u}^m, p^m)$, one should solve, when $\theta \neq 0$, the equivalent problem

$$\frac{1}{\delta\theta}(\mathbf{u}_\theta^m, \mathbf{v}^h) - \mathbf{F}(\mathbf{f}_\theta^m, \mathbf{u}_\theta^m, p_\theta^m; \mathbf{v}^h) = \frac{1}{\delta\theta}(\mathbf{u}^{m-1}, \mathbf{v}^h) \qquad \text{for all } \mathbf{v}^h \in \mathbf{V}_0^h \quad (10.4)$$

and

$$b(\mathbf{u}_\theta^m, q^h) = \begin{cases} (1 - \theta)b(\mathbf{u}^0, q^h) & \text{if } m = 1 \\ 0 & \text{if } m > 1 \end{cases} \qquad \text{for all } q^h \in S_0^h \quad (10.5)$$

and then set

$$\mathbf{u}^m = \frac{1}{\theta}\mathbf{u}_\theta^m - \frac{1-\theta}{\theta}\mathbf{u}^{m-1} \qquad \text{and} \qquad p^m = \frac{1}{\theta}p_\theta^m - \frac{1-\theta}{\theta}p^{m-1}. \quad (10.6)$$

Substituting (9.15) into (10.4) yields

$$a(\mathbf{u}_\theta^m, \mathbf{v}^h) + c(\mathbf{u}_\theta^m, \mathbf{u}_\theta^m, \mathbf{v}^h) + b(\mathbf{v}^h, p_\theta^m) + \frac{1}{\delta\theta}(\mathbf{u}_\theta^m, \mathbf{v}^h)$$

$$= (\mathbf{f}_\theta^m, \mathbf{v}^h) + \frac{1}{\delta\theta}(\mathbf{u}^{m-1}, \mathbf{v}^h) \qquad \text{for all } \mathbf{v}^h \in \mathbf{V}_0^h. \quad (10.7)$$

The system of algebraic equations (10.5) and (10.7) is very similar to that encountered in the stationary case, i.e., compare with (2.1)–(2.2). Indeed, the only differences are the last terms on both sides of (10.7). In principle one could solve (10.5) and (10.7) by any of the methods discussed in Chapter 5. Note that if δ is small, then a good starting guess for any iterative method for solving (10.5) and (10.7) is the solution $\mathbf{u}^{m-1}$ at the previous time step. On the other hand, due to the fully implicit character of the scheme (10.1)–(10.3) whenever $\theta \neq 0$, a different nonlinear system has to be solved for each m.

As has been noted, the above scheme is first-order accurate with respect to δ whenever $\theta \neq 1/2$ and is second-order accurate, i.e., $O(\delta^2)$, for $\theta = 1/2$. These observations are with respect to both the $\mathbf{L}^2(\Omega)$ and $\mathbf{H}_0^1(\Omega)$ norms of the differences $\mathbf{u}(m\delta, \cdot) - \mathbf{u}^m$, $m = 1, ..., M$.

10.2. Semi-Implicit Single-Step Methods

The fully implicit methods discussed in Section 9.1 require the solution of a nonlinear system at every time step. One way to view a semi-implicit method is as a linearization of some fully implicit method. For example, the scheme (10.1)–(10.3) may be linearized by Newton's method and it becomes a semi-implicit method if we additionally decide to perform only one iteration at each time step. Then, using the economical form (10.7) along with (10.5), we would compute, given $\mathbf{u}^{m-1}$, the pair $(\mathbf{u}_\theta^m, p_\theta^m)$ by solving the linear system

$$a(\mathbf{u}_\theta^m, \mathbf{v}^h) + c(\mathbf{u}^{m-1}, \mathbf{u}_\theta^m, \mathbf{v}^h) + c(\mathbf{u}_\theta^m, \mathbf{u}^{m-1}, \mathbf{v}^h) + b(\mathbf{v}^h, p_\theta^m) + \frac{1}{\delta\theta}(\mathbf{u}_\theta^m, \mathbf{v}^h)$$

$$= (\mathbf{f}_\theta^m, \mathbf{v}^h) + c(\mathbf{u}^{m-1}, \mathbf{u}^{m-1}, \mathbf{v}^h) + \frac{1}{\delta\theta}(\mathbf{u}^{m-1}, \mathbf{v}^h) \quad \text{for all } \mathbf{v}^h \in \mathbf{V}_0^h$$

and   (10.8)

$$b(\mathbf{u}_\theta^m, q^h) = \begin{cases} (1 - \theta)b(\mathbf{u}^0, q^h) & \text{if } m = 1 \\ 0 & \text{if } m > 1 \end{cases} \quad \text{for all } q^h \in S_0^h. \quad (10.9)$$

Again, $(\mathbf{u}^m, p^m)$ may be recovered from $(\mathbf{u}_\theta^m, p_\theta^m)$ by (10.6). Note that the single linear system that must be solved at every time step changes from time step to time step.

A simpler linearization of (10.5) and (10.7) results by applying the method of Section 6.3; it is given by

$$a(\mathbf{u}_\theta^m, \mathbf{v}^h) + c(\mathbf{u}^{m-1}, \mathbf{u}_\theta^m, \mathbf{v}^h) + b(\mathbf{v}^h, p_\theta^m) + \frac{1}{\delta\theta}(\mathbf{u}_\theta^m, \mathbf{v}^h)$$

$$= (\mathbf{f}_\theta^m, \mathbf{v}^h) + \frac{1}{\delta\theta}(\mathbf{u}^{m-1}, \mathbf{v}^h) \quad \text{for all } \mathbf{v}^h \in \mathbf{V}_0^h \quad (10.10)$$

and

$$b(\mathbf{u}_\theta^m, q^h) = \begin{cases} (1 - \theta)b(\mathbf{u}^0, q^h) & \text{if } m = 1 \\ 0 & \text{if } m > 1 \end{cases} \quad \text{for all } q^h \in S_0^h. \quad (10.11)$$

This last scheme has been analyzed in Temam [1979], where the unconditional stability for $\theta = 1$ and $\theta = 1/2$ is proved. Also, the order of accuracy of these schemes are similar to that of their parent scheme (10.5)–(10.7).

10.3. Backward Differentiation Multistep Methods

The multistep schemes we consider are of the backward differentiation type, of which the backward Euler method is the simplest prototype. One advantage of these methods is that they do not require an initial guess for the pressure. The k-step method has the following form.

Given $\mathbf{u}^0, \mathbf{u}^1, \ldots, \mathbf{u}^{k-1}, \{\mathbf{u}^m, p^m\}$ for $m = k, \ldots, M$ are determined from

$$\frac{1}{\delta}(D_k \mathbf{u}^m, \mathbf{v}^h) = \mathbf{F}(\mathbf{f}^m, \mathbf{u}^m, p^m; \mathbf{v}^h) \quad \text{for all } \mathbf{v}^h \in \mathbf{V}_0^h \quad (10.12)$$

and

$$b(\mathbf{u}^m, q^h) = 0 \quad \text{for all } q^h \in S_0^h, \quad (10.13)$$

where D_k is a difference operator. For $k = 1, 2,$ and $3,$ D_k is given by

$$D_1 \mathbf{u}^m = \mathbf{u}^m - \mathbf{u}^{m-1},$$

$$D_2 \mathbf{u}^m = \frac{3\mathbf{u}^m - 4\mathbf{u}^{m-1} + \mathbf{u}^{m-2}}{2},$$

and

$$D_3 \mathbf{u}^m = \frac{11\mathbf{u}^m - 18\mathbf{u}^{m-1} + 9\mathbf{u}^{m-2} - 2\mathbf{u}^{m-3}}{6}.$$

Using (9.15), (10.12)–(10.13) are equivalent to, for $m = k, \ldots, M,$

$$a(\mathbf{u}^m, \mathbf{v}^h) + c(\mathbf{u}^m, \mathbf{u}^m, \mathbf{v}^h) + b(\mathbf{v}^h, p^m) + \frac{c_{kk}}{\delta}(\mathbf{u}^m, \mathbf{v}^h)$$

$$= (\mathbf{f}^m, \mathbf{v}^h) - \frac{1}{\delta}\left(\sum_{j=0}^{k-1} c_{kj} u^{m+j-k}, \mathbf{v}^h\right) \quad \text{for all } \mathbf{v}^h \in \mathbf{V}_0^h \quad (10.14)$$

and

$$b(\mathbf{u}^m, q^h) = 0 \quad \text{for all } q^h \in S_0^h, \quad (10.15)$$

where

$$c_{11} = 1, \quad c_{10} = -1,$$

$$c_{22} = \frac{3}{2}, \quad c_{21} = -2, \quad c_{20} = \frac{1}{2}, \quad (10.16)$$

$$c_{33} = \frac{11}{6}, \quad c_{32} = -3, \quad c_{31} = \frac{3}{2}, \quad c_{30} = -\frac{1}{3}.$$

Again, the nonlinear system (10.14)–(10.15) is very similar to that encountered in the stationary case. Also, at every time step a different nonlinear system must be solved.

One disadvantage of multistep methods is that they require more starting values than just $\mathbf{u}^0$. Here $\mathbf{u}^0$ may be chosen to be $\mathbf{u}_0$. For the two-step scheme, i.e., (10.14)–(10.15) with $k = 2$, $\mathbf{u}^1$ may be obtained using a single step of the one-step backward Euler scheme, i.e., (10.14)–(10.15) with $k = 1$. This does not deteriorate the accuracy of the two-step scheme since the local accuracy, i.e., the accuracy over one step, of the backward Euler method is presumably the same as the global accuracy, i.e., the accuracy over many steps, of the two-step method. For the three-step scheme, i.e., (10.14)–(10.15) with $k = 3$, in order to preserve the higher accuracy, we solve for $\mathbf{u}^1$ using the Crank–Nicolson method, and then $\mathbf{u}^2$ is found from $\mathbf{u}^1$ and $\mathbf{u}^0$ by using the two-step scheme.

These fully implicit multistep schemes are unconditionally stable, and, at least for the velocity field, are accurate of $O(\delta^k)$. One may effect linearizations similar to those discussed in Section 10.2, yielding methods that require the solution of a single differing linear system at every time step. These types of methods are not very popular because the cost of the latter task is rather onerous.

10.4. A Class of Semi-Implicit Multistep Methods

We now consider semi-implicit methods that have the feature that at each time step a single linear system has to be solved, and furthermore, that this linear system is the *same* for every time step. Thus, in order to compute the sequence $\{\mathbf{u}^m, p^m\}$, a *single* matrix has to be factored. In order to accomplish this, the nonlinear terms in the discrete equations have to be completely lagged, i.e., evaluated using only information from previous time steps. The price to be paid for this substantial simplification is that these methods are only conditionally stable. However, their stability restrictions are determined by the convection speeds and are not especially onerous. In fact, accuracy considerations would probably require that the time step be chosen in a similar manner to that ruled by the stability restriction.

We will focus on semi-implicit variants of the schemes discussed in Section 10.3. Our discussion follows along the lines of Baker, Dougalis, and Karakashian [1982]. Again, these methods do not require an initial guess for the pressure. The k-step method has the following form. Given $\mathbf{u}^0, \mathbf{u}^1, ..., \mathbf{u}^{k-1}$, $\{\mathbf{u}^m, p^m\}$ for $m = k, ..., M$ are determined from

$$a(\mathbf{u}^m, \mathbf{v}^h) + b(\mathbf{v}^h, p^m) + \frac{c_{kk}}{\delta}(\mathbf{u}^m, \mathbf{v}^h) = (\mathbf{f}^m, \mathbf{v}^h) - c(\Lambda_k \mathbf{u}^m, \Lambda_k \mathbf{u}^m, \mathbf{v}^h)$$

$$-\frac{1}{\delta}\sum_{j=0}^{k-1} c_{kj}(\mathbf{u}^{m+j-k}, \mathbf{v}^h) \quad \text{for all } \mathbf{v}^h \in \mathbf{V}_0^h \tag{10.17}$$

and

$$b(\mathbf{u}^m, q^h) = 0 \quad \text{for all } q^h \in S_0^h, \tag{10.18}$$

where the c_{kj}'s are given in (10.16) and where the lagging difference operators Λ_k for $k = 1, 2, 3$ are given by

$$\Lambda_1 \mathbf{u}^m = \mathbf{u}^{m-1},$$

$$\Lambda_2 \mathbf{u}^m = 2\mathbf{u}^{m-1} - \mathbf{u}^{m-2},$$

and

$$\Lambda_3 \mathbf{u}^m = 3\mathbf{u}^{m-1} - 3\mathbf{u}^{m-2} + \mathbf{u}^{m-3}.$$

Clearly, (10.17)–(10.18) constitute a linear system for the pair $(\mathbf{u}^m, p^m)$, and it is easy to see that the coefficient matrix for this linear system is independent of m, i.e., is the same at each time step.

For $k > 1$, (10.17)–(10.18) requires starting conditions above just $\mathbf{u}^0$. We would like to generate these using the same linear system that appears in (10.17)–(10.18) so that the whole computation can be accomplished with only a single matrix factorization. For $k = 2$, this can be accomplished, without deteriorating the accuracy, by the following procedure. First, $\mathbf{u}^0$ may be set to $\mathbf{u}_0$. We then perform two steps of the semi-implicit backward Euler method, i.e., (10.17)–(10.18) with $k = 1$, using a time step of $2\delta/3$. Thus we define the intermediate values $(\mathbf{u}^{2/3}, p^{2/3})$ by

$$a(\mathbf{u}^{2/3}, \mathbf{v}^h) + b(\mathbf{v}^h, p^{2/3}) + \frac{3}{2\delta}(\mathbf{u}^{2/3}, \mathbf{v}^h)$$

$$= (\mathbf{f}^{2/3}, \mathbf{v}^h) - c(\mathbf{u}^0, \mathbf{u}^0, \mathbf{v}^h) - \frac{3}{2\delta}(\mathbf{u}^0, \mathbf{v}^h) \quad \text{for all } \mathbf{v}^h \in \mathbf{V}_0^h \tag{10.19}$$

and

$$b(\mathbf{u}^{2/3}, q^h) = 0 \quad \text{for all } q^h \in S_0^h, \tag{10.20}$$

and $(\mathbf{u}^{4/3}, p^{4/3})$ by

$$a(\mathbf{u}^{4/3}, \mathbf{v}^h) + b(\mathbf{v}^h, p^{4/3}) + \frac{3}{2\delta}(\mathbf{u}^{4/3}, \mathbf{v}^h)$$

$$= (\mathbf{f}^{4/3}, \mathbf{v}^h) - c(\mathbf{u}^{2/3}, \mathbf{u}^{2/3}, \mathbf{v}^h) - \frac{3}{2\delta}(\mathbf{u}^{2/3}, \mathbf{v}^h) \quad \text{for all } \mathbf{v}^h \in \mathbf{V}_0^h \tag{10.21}$$

and

$$b(\mathbf{u}^{4/3}, q^h) = 0 \quad \text{for all } q^h \in S_0^h.$$ (10.22)

Then, we define

$$\mathbf{u}^1 = \frac{\mathbf{u}^{4/3} + \mathbf{u}^{2/3}}{2}.$$ (10.23)

Note that the linear systems that need to be solved in (10.19)–(10.20) and (10.21)–(10.22) are identical to the one appearing in (10.17)–(10.18) with $k = 2$. To summarize, the complete two-step algorithm is given by (10.17)–(10.18) with $k = 2$ and (10.19)–(10.23).

The procedure for generating $\mathbf{u}^0$, $\mathbf{u}^1$, and $\mathbf{u}^2$ for the case $k = 3$ is substantially more complicated. However, by solving a sequence of five linear systems, all with the same coefficient matrix as that appearing in (10.17)–(10.18) with $k = 3$, starting values $\mathbf{u}^0$, $\mathbf{u}^1$, and $\mathbf{u}^2$ may be found that do not deteriorate the overall accuracy of the method. We denote by $L_3(\cdot, \cdot)$ the linear operator on $(\mathbf{u}^m, p^m)$ appearing on the left-hand side of (10.17) with $k = 3$ so that

$$L_3(\mathbf{u}, p) = a(\mathbf{u}, \mathbf{v}^h) + b(\mathbf{v}^h, p) + \frac{11}{6\delta}(\mathbf{u}, \mathbf{v}^h).$$

We first determine $\mathbf{u}^0$ and p^0, the latter not subsequently needed, by solving

$$L_3(\mathbf{u}^0, p^0) = \frac{11}{6\delta}(\mathbf{u}_0, \mathbf{v}^h) + a(\mathbf{u}_0, \mathbf{v}^h) \quad \text{for all } \mathbf{v}^h \in \mathbf{V}_0^h$$ (10.24)

and

$$b(\mathbf{u}^0, q^h) = 0 \quad \text{for all } q^h \in S_0^h,$$ (10.25)

where $\mathbf{u}_0$ is the given initial condition. Thus, here the first starting value $\mathbf{u}^0$ is determined through a specific projection. Next the three intermediate pairs $(\tilde{\mathbf{u}}^j, \tilde{p}^j)$, $j = 1, 2, 3$ are determined from the linear systems

$$L_3(\tilde{\mathbf{u}}^1, \tilde{p}^1) = \frac{11}{6\delta}(\mathbf{u}^0, \mathbf{v}^h) + (\mathbf{f}^{6/11}, \mathbf{v}^h)$$

$$- c(\mathbf{u}^0, \mathbf{u}^0, \mathbf{v}^h) \quad \text{for all } \mathbf{v}^h \in \mathbf{V}_0^h$$ (10.26)

and

$$b(\tilde{\mathbf{u}}^1, q^h) = 0 \quad \text{for all } q^h \in S_0^h,$$ (10.27)

defining $(\tilde{\mathbf{u}}^1, \tilde{p}^1)$,

$$L_3(\tilde{\mathbf{u}}^2, \tilde{p}^2) = \frac{11}{6\delta}(\mathbf{u}^0, \mathbf{v}^h) + (\mathbf{f}^{12/11} + \mathbf{f}^0, \mathbf{v}^h) - c(\mathbf{u}^0, \mathbf{u}^0, \mathbf{v}^h)$$

$$- c(2\tilde{\mathbf{u}}^1 - \mathbf{u}^0, 2\tilde{\mathbf{u}}^1 - \mathbf{u}^0, \mathbf{v}^h) \quad \text{for all } \mathbf{v}^h \in \mathbf{V}_0^h, \quad (10.28)$$

and

$$b(\tilde{\mathbf{u}}^2, q^h) = 0 \quad \text{for all } q^h \in S_0^h, \quad (10.29)$$

defining $(\tilde{\mathbf{u}}^2, \tilde{p}^2)$, and

$$L_3(\tilde{\mathbf{u}}^3, \tilde{p}^3) = \frac{11}{6\delta}(\mathbf{u}^0, \mathbf{v}^h) + (\mathbf{f}^{12/11} + \mathbf{f}^0, \mathbf{v}^h) - c(\mathbf{u}^0, \mathbf{u}^0, \mathbf{v}^h)$$

$$- c(2\tilde{\mathbf{u}}^2 - \mathbf{u}^0, 2\tilde{\mathbf{u}}^2 - \mathbf{u}^0, \mathbf{v}^h) \quad \text{for all } \mathbf{v}^h \in \mathbf{V}_0^h, \quad (10.30)$$

and

$$b(\tilde{\mathbf{u}}^3, q^h) = 0 \quad \text{for all } q^h \in S_0^h, \quad (10.31)$$

defining $(\tilde{\mathbf{u}}^3, \tilde{p}^3)$. Then, we set

$$\mathbf{u}^{12/11} = 2\tilde{\mathbf{u}}^3 - \mathbf{u}^0. \quad (10.32)$$

The fifth linear system to be solved is

$$L_3(\tilde{\mathbf{u}}^4, \tilde{p}^4) = \frac{11}{6\delta}(\mathbf{u}^{12/11}, \mathbf{v}^h) + (\mathbf{f}^{24/11} + \mathbf{f}^{12/11}, \mathbf{v}^h) - c(\mathbf{u}^{12/11}, \mathbf{u}^{12/11}, \mathbf{v}^h)$$

$$- c(2\mathbf{u}^{12/11} - \mathbf{u}^0, 2\mathbf{u}^{12/11} - \mathbf{u}^0, \mathbf{v}^h) \quad \text{for all } \mathbf{v}^h \in \mathbf{V}_0^h$$

$$(10.33)$$

and

$$b(\tilde{\mathbf{u}}^4, q^h) = 0 \quad \text{for all } q^h \in S_0^h \quad (10.34)$$

for $(\tilde{\mathbf{u}}^4, \tilde{p}^4)$. Then, we set

$$\mathbf{u}^{24/11} = 2\tilde{\mathbf{u}}^4 - \mathbf{u}^{12/11}. \quad (10.35)$$

Finally, the starting conditions $\mathbf{u}^1$ and $\mathbf{u}^2$ are defined by

$$\mathbf{u}^1 = \frac{13\mathbf{u}^0 + 286\mathbf{u}^{12/11} - 11\mathbf{u}^{24/11}}{288} \quad (10.36)$$

and

$$\mathbf{u}^2 = \frac{-5\mathbf{u}^0 + 22\mathbf{u}^{12/11} + 55\mathbf{u}^{24/11}}{72}. \quad (10.37)$$

To summarize, the complete $k = 3$ method is given by (10.17)–(10.18) with $k = 3$ and (10.24)–(10.37).

These multistep methods are only conditionally stable. For example, in a slightly different context, it can be shown that these multistep methods are stable for sufficiently small h and δ satisfying $\delta \le Ch^s$, where $s = 1, 4/5, 4/7$ for $k = 1, 2, 3$, respectively. These methods are known to be optimally accurate, i.e., for $k = 1, 2, 3$, we have that the velocity errors, in either the $\mathbf{L}^2(\Omega)$ or $\mathbf{H}_0^1(\Omega)$-norms, are $O(\delta^k)$ whenever the solution of (9.1)–(9.4) is sufficiently smooth. See Baker, Dougalis, and Karakashian [1982] for details.

In our own experience, we have found the $k = 2$ method of this section to provide the best balance between accuracy and complexity.

IV

THE STREAMFUNCTION-VORTICITY FORMULATION

In the past, difficulties associated with the stable approximation of the incompressibility constraint made it desirable to use alternate formulations of the equations of viscous flows wherein the incompressibility constraint does not explicitly appear. The most commonly used alternative for plane or axially symmetric flows has been the *streamfunction-vorticity* formulation. Another advantage, in two-dimensional settings, is that it involves only two unknown scalar fields, namely, the streamfunction and vorticity, in contrast to the primitive variable formulation using velocity and pressure. However, for flows in multiply connected regions, the streamfunction-vorticity formulation suffers the disadvantage of having unknown boundary information that must be determined as part of the solution.

Here we discuss the finite element approximation of the streamfunction-vorticity equations, concentrating on the case of *plane flows*. In addition to defining weak formulations, listing useful finite element spaces, and discussing error estimates, we are interested in various methods for dealing with flows in *multiply connected* regions. Also of interest is the recovery of approximations to the primitive variables from the approximate streamfunction and vorticity fields. We will only touch on this recovery procedure since it parallels the analogous one of Section 13.4 connected with the streamfunction formulation.

We do not consider in detail three-dimensional formulations involving the streamfunctions and/or the vorticity field such as the velocity-vorticity method or methods using the velocity potential. Presently, this is a subject of very intense research and of substantial promise; see the brief discussion in Part X.

11

Algorithms for the Streamfunction-Vorticity Equations

Before beginning our examination of streamfunction-vorticity formulations, we note that in $\mathbb{R}^2$ one must differentiate between two different curl operators. The first takes vectors into scalars while the other does the opposite. Thus, in $\mathbb{R}^2$, we define the *vorticity* ω to be the scalar field

$$\omega = \text{curl } \mathbf{u} = \frac{\partial u_2}{\partial x_1} - \frac{\partial u_1}{\partial x_2}, \tag{11.1}$$

where $\mathbf{u} = (u_1, u_2)^T$ is the velocity field, while we define the *streamfunction* ψ to be a scalar field such that

$$\mathbf{u} = \text{curl } \psi = \left(\frac{\partial \psi}{\partial x_2}, -\frac{\partial \psi}{\partial x_1} \right)^T. \tag{11.2}$$

Since it will always be clear from the context which of the two curl operators is being used, we use the same notation for both. It is well known that these two operators are formal adjoint operators in the sense that

$$\int_\Omega \mathbf{v} \cdot \text{curl } \phi \, d\Omega - \int_\Omega \phi \, \text{curl } \mathbf{v} \, d\Omega = \int_\Gamma \phi \mathbf{v} \times \mathbf{n} \, d\Gamma$$

whenever the relevant operations are justified and where Γ denotes the boundary of $\Omega \subset \mathbb{R}^2$. Thus, in $\mathbb{R}^2$, the pair of curl operators enjoy a relationship analogous to that of the pair (div, -grad). Of course, for any sufficiently smooth scalar function ϕ, we also have the identities div(curl ϕ) = 0 and curl(grad ϕ) = 0.

Through the introduction of the streamfunction ψ satisfying (11.2), the incompressibility constraint (1.2) is automatically satisfied. Combining (11.1) and (11.2) yields

$$-\Delta\psi = \omega. \tag{11.3}$$

Taking the curl of the momentum equation (1.1) yields, with the aid of the continuity equation,

$$-\nu\,\Delta\omega = -\text{curl }\psi \cdot \text{grad }\omega + \text{curl f}. \tag{11.4}$$

We suppose that these equations hold in an open, bounded, possibly multiply connected domain $\Omega \subset \mathbb{R}^2$. Of course, ν and $\mathbf{f}$ retain their meanings from the primitive variable case. However, we note that (11.4) may be derived from (1.1) and (1.2) only when ν = constant.

11.1. Boundary Conditions for the Streamfunction-Vorticity Equations

We denote the boundary of Ω by Γ and assume that it is composed of closed, nonintersecting segments Γ_i, $i = 1, ..., m$, enclosed by the closed curve Γ_0. The exterior boundary Γ_0 may be a true physical boundary, it may be an artificial boundary introduced in order to render the computational domain finite, or it may be a combination of both. Furthermore, we assume that Γ_0 may be divided into three portions Γ_{0j}, $j = 1, 2, 3$, in such a way that $\bar{\Gamma}_0 = \bigcup_{j=1}^{3} \bar{\Gamma}_{0j}$ and $\Gamma_{0j} \cap \Gamma_{0k} = 0$ for $j \neq k$. See Figure 11.1 for an example. Each portion Γ_{0j} may itself consist of disjoint segments and as many as two of them may be empty. We will have to differentiate between the two cases

Case 1. $\Gamma_{01} \cup \Gamma_{02}$ is not empty

and

Case 2. $\Gamma_{01} \cup \Gamma_{02}$ is empty.

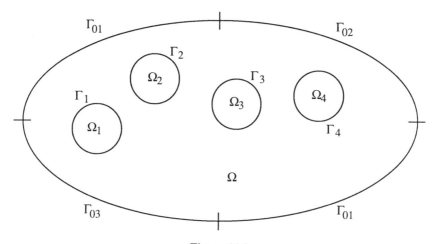

Figure 11.1
A multiply connected domain.

Solely to simplify the exposition, we assume that in Case 1 $\Gamma_{01} \cup \Gamma_{02}$ forms a connected subset of Γ_0, although Γ_{01} and Γ_{02} themselves may consist of disjoint segments.

The boundary conditions that are to be imposed are given by

$$\mathbf{u} = \mathbf{g}_i \text{ on } \Gamma_i \quad \text{for } i = 1, \ldots, m, \tag{11.5}$$

$$\mathbf{u} = \mathbf{g}_0 \text{ on } \Gamma_{01}, \tag{11.6}$$

$$\mathbf{u} \cdot \mathbf{n} = g_n \quad \text{and} \quad \omega = \omega_0 \text{ on } \Gamma_{02}, \tag{11.7}$$

and

$$\mathbf{u} \cdot \mathbf{\tau} = g_\tau \quad \text{and} \quad \frac{\partial \omega}{\partial n} = \omega_n \text{ on } \Gamma_{03}, \tag{11.8}$$

where $\mathbf{n}$ and τ denote the unit outer normal and counterclockwise tangent vectors to the boundary. These boundary conditions on the velocity and vorticity are of use in a variety of physical and computational settings; see Gunzburger and Peterson [1988] for details. From (11.6)–(11.7) we see that Case 1 corresponds to the normal component of the velocity being specified on at least part of the exterior boundary Γ_0; for Case 2 the normal velocity is specified nowhere on the exterior boundary Γ_0. We note that a streamfunction exists only if

$$\int_{\Gamma_i} \mathbf{g}_i \cdot \mathbf{n} \, d\Gamma = 0 \quad \text{for } i = 1, \ldots, m$$

and

$$\int_{\Gamma_{01}} \mathbf{g}_0 \cdot \mathbf{n} \, d\Gamma + \int_{\Gamma_{02}} g_n \, d\Gamma = 0 \qquad \text{if } \Gamma_0 = \Gamma_{01} \cup \Gamma_{02}.$$

These simply state that no net mass flows through any of the boundary pieces. Throughout we will assume that the given data $\mathbf{g}_i$, $i = 0, ..., m$, satisfies these relations.

In terms of the streamfunction and vorticity, (11.5)–(11.8) imply that

$$\psi = q_i + a_i \quad \text{and} \quad \frac{\partial \psi}{\partial n} = -\mathbf{g}_i \cdot \tau \text{ on } \Gamma_i \qquad \text{for } i = 1, ..., m, \quad (11.9)$$

$$\psi = q_0 \quad \text{and} \quad \frac{\partial \psi}{\partial n} = -\mathbf{g}_0 \cdot \tau \text{ on } \Gamma_{01}, \qquad\qquad (11.10)$$

$$\psi = q_0 \quad \text{and} \quad \omega = \omega_0 \text{ on } \Gamma_{02}, \qquad\qquad (11.11)$$

and

$$\frac{\partial \psi}{\partial n} = -g_\tau \quad \text{and} \quad \frac{\partial \omega}{\partial n} = \omega_n \text{ on } \Gamma_{03}. \qquad\qquad (11.12)$$

In (11.9), $a_m = 0$ for Case 2 and in (11.9)–(11.11) q_i, $i = 0, ..., m$, denote functions such that

$$\frac{\partial q_i}{\partial \tau} = \mathbf{g}_i \cdot \mathbf{n} \text{ on } \Gamma_i \qquad \text{for } i = 1, ..., m,$$

$$q_m(\mathbf{x}_0) = 0 \qquad \text{for some } \mathbf{x}_0 \in \Gamma_m \qquad \text{for Case 2,}$$

$$\frac{\partial q_0}{\partial \tau} = \mathbf{g}_0 \cdot \mathbf{n} \text{ on } \Gamma_{01}, \qquad \frac{\partial q_0}{\partial \tau} = g_n \text{ on } \Gamma_{02},$$

and

$$q_0(\mathbf{x}_0) = 0 \qquad \text{for some } \mathbf{x}_0 \in \Gamma_{01} \cup \Gamma_{02} \text{ for Case 1.}$$

These functions may always be defined in terms of integrals of $\mathbf{g}_i \cdot \mathbf{n}$ or g_n, as the case may be, along the corresponding boundary segments. These integrals can often be explicitly evaluated or, in any case, can be evaluated to arbitrary accuracy through numerical procedures.

The arbitrary numbers a_i, $i = 1, ..., m$, appearing in (11.9) result from the well-known fact that the streamfunction is uniquely determined up to a *single* additive constant. We have fixed that constant at an arbitrary point on $\Gamma_{01} \cup \Gamma_{02}$, or, if these segments are both empty, at an arbitrary point on Γ_m. Then a_i gives a measure of the unknown mass flow across any curve joining Γ_i and the boundary segment at which the streamfunction has been fixed, i.e., either $\Gamma_{01} \cup \Gamma_{02}$ or Γ_m.

Due to the appearance of the additional unknowns a_i, additional data must be supplied. This data, which may be derived from the requirement that the pressure be a single-valued function, is given by (Girault and Raviart [1979 and 1986]; Gunzburger and Peterson [1988]; and Tezduyar, Glowinski, and Liou [1988])

$$\int_{\Gamma_i} \left(\nu \frac{\partial \omega}{\partial n} - \omega \mathbf{g}_i \cdot \mathbf{n} + \mathbf{f} \cdot \boldsymbol{\tau} \right) d\Gamma = 0 \qquad \text{for } i = 1, \ldots, m_1, \quad (11.13)$$

where $m_1 = m$ for Case 1 and $m_1 = m - 1$ for Case 2.

Thus the general streamfunction-vorticity problem we consider is to find ψ, ω, and a_i, $i = 1, \ldots, m_1$, such that (11.3), (11.4), and (11.9)–(11.13) are satisfied. As is usually the case, finite element algorithms are defined through weak formulations. Before introducing the latter we will need to discuss some additional function spaces.

11.2. Function Spaces and Forms

In order to give a precise definition of a weak formulation on which finite element discretizations are to be based, we need to introduce some formidable looking function spaces. Fortunately, if one ignores these function analytic niceties, the material of this chapter may still be understood from an algorithmic viewpoint. Those who wish to delve more deeply into details concerning these function spaces may consult, e.g., Adams [1975] and Grisvard [1985].

The function spaces defined in Section 1.1 are not by themselves adequate to define meaningful weak formulations of the streamfunction-vorticity equations. Therefore, we here introduce, for integers $r \geq 0$ and $p \geq 1$, the Sobolev spaces $W^{r,p}(\Omega)$ consisting of functions having derivatives of order up to r belonging to $L^p(\Omega)$; $\| \cdot \|_{r,p}$ will denote the usual norm for functions belonging to $W^{r,p}(\Omega)$. Of course, $W^{r,2}(\Omega) = H^r(\Omega)$ and $W^{0,p}(\Omega) = L^p(\Omega)$. We will be particularly interested in the cases $r = 0$ and 1 and $p = 2$ and 4.

Next we define the sets

$$V = \{ \zeta \in W^{1,2}(\Omega) : \zeta = \omega_0 \text{ on } \Gamma_{02} \},$$

and

$$S = \{\phi \in W^{1,4}(\Omega): \phi = q_0 \text{ on } \Gamma_{01} \cup \Gamma_{02}; \quad \phi = q_i + a_i \text{ on } \Gamma_i, \, i = 1, \ldots, m,$$

$$a_i \in \mathbb{R} \text{ arbitrary except that } a_m = 0 \text{ for Case 2}\}.$$

and the spaces

$$V_0 = \{\zeta \in W^{1,2}(\Omega): \zeta = 0 \text{ on } \Gamma_{02}\}$$

and

$$S_0 = \{\phi \in W^{1,4}(\Omega): \phi = 0 \text{ on } \Gamma_{01} \cup \Gamma_{02}; \quad \phi = a_i \text{ on } \Gamma_i, \, i = 1, \ldots, m,$$

$$a_i \in \mathbb{R} \text{ arbitrary except that } a_m = 0 \text{ for Case 2}\}.$$

We will use the spaces $W^{(p-1)/p,p}(\tilde{\Gamma})$ consisting of traces of functions belonging to $W^{1,p}(\Omega)$, where $\tilde{\Gamma}$ denotes an appropriate boundary segment. A norm for functions $q \in W^{(p-1)/p,p}(\tilde{\Gamma})$ is given by

$$\|q\|_{(p-1)/p,p,\tilde{\Gamma}} = \inf_{\substack{v \in W^{1,p}(\Omega) \\ v|_{\tilde{\Gamma}} = q}} \|v\|_{1,p}.$$

Also, we will use the dual spaces with negative indices of some of the spaces defined above. For example, we will use the space $W^{(1-p)/p,p/(p-1)}(\tilde{\Gamma})$, which is dual to $W^{(p-1)/p,p}(\tilde{\Gamma})$ and is provided with the norm

$$\|q\|_{(1-p)/p,p/(p-1),\tilde{\Gamma}} = \sup_{0 \neq v \in W^{(p-1)/p,p}(\tilde{\Gamma})} \frac{\int_{\tilde{\Gamma}} qv \, d\Gamma}{\|v\|_{(p-1)/p,p,\tilde{\Gamma}}}.$$

Spaces of vector valued functions are defined from the spaces of scalar valued functions in the usual manner. For details concerning any of these spaces, see Adams [1975] and Grisvard [1985], especially with regard to the required smoothness of the boundary needed to make the above definitions and subsequent results valid.

Having defined the above function spaces, we now define the bilinear forms

$$A(\omega, \zeta) = \int_\Omega \omega \zeta \, d\Omega \quad \text{for all } \omega, \zeta \in W^{1,2}(\Omega) = H^1(\Omega) \quad (11.14)$$

and

$$B(\zeta, \phi) = -\int_\Omega \text{grad } \zeta \cdot \text{grad } \phi \, d\Omega$$

$$\text{for all } \zeta \in W^{1,2}(\Omega) \text{ and } \phi \in W^{1,4}(\Omega), \quad (11.15)$$

and the trilinear form

$$C(\zeta,\, \psi,\, \phi) = -\int_{\Omega} \zeta \operatorname{grad} \phi \cdot \operatorname{curl} \psi \, d\Omega - \int_{\Gamma_{03}} \zeta\phi \frac{\partial\psi}{\partial\tau} \, d\Gamma$$

$$\text{for all } \zeta \in W^{1,2}(\Omega) \text{ and } \psi,\, \phi \in W^{1,4}(\Omega). \tag{11.16}$$

It is the presence of the trilinear form (11.16) that motivates the use of the space $W^{1,4}(\Omega)$. Indeed, (11.16) is not well defined if ψ and ϕ merely belong to $H^1(\Omega) = W^{1,2}(\Omega)$. Note that (11.16) does not enter weak formulations of the *linear* Stokes problem, and thus for this problem, one may work with the space $H^1(\Omega) = W^{1,2}(\Omega)$ for both the stream-function and vorticity.

11.3. A Weak Formulation and Its Associated Natural Boundary Conditions

The weak formulation of the streamfunction-vorticity equations that we will examine is defined as follows. Given $\mathbf{f} \in \mathbf{L}^4(\Omega)$, $\mathbf{g}_0 \in \mathbf{W}^{-3/4,4/3}(\Gamma_{01})$, $\mathbf{g}_i \in \mathbf{W}^{-3/4,4/3}(\Gamma_i)$ for $i = 1, \dots, m$, $g_n \in \mathbf{W}^{-3/4,4/3}(\Gamma_{02})$, $g_\tau \in \mathbf{W}^{-3/4,4/3}(\Gamma_{03})$, $\omega_0 \in W^{1/2,2}(\Gamma_{02})$, and $\omega_n \in W^{-1/2,2}(\Gamma_{03})$, seek $\psi \in V$, $\omega \in S$, and $a_i \in \mathbb{R}$ for $i = 1, \dots, m_1$ such that

$$A(\omega,\, \zeta) + B(\zeta,\, \psi) = F(\zeta) \qquad \text{for all } \zeta \in V_0 \tag{11.17}$$

and

$$vB(\omega,\, \phi) + C(\omega,\, \psi,\, \phi) = G(\phi) \qquad \text{for all } \phi \in S_0. \tag{11.18}$$

The linear functionals $F(\cdot)$ and $G(\cdot)$ are given by

$$F(\zeta) = \int_{\Gamma_{01}} \zeta\mathbf{g}_0 \cdot \tau \, d\Gamma + \int_{\Gamma_{03}} \zeta g_\tau \, d\Gamma + \sum_{i=1}^{m} \int_{\Gamma_i} \zeta\mathbf{g}_i \cdot \tau \, d\Gamma \qquad \text{for all } \zeta \in V_0$$

and

$$G(\phi) = -\int_{\Omega} \phi \operatorname{curl} \mathbf{f} \, d\Gamma - v \int_{\Gamma_{03}} \phi\omega_n \, d\Gamma$$

$$+ \sum_{i=1}^{m_1} \int_{\Gamma_i} \phi\mathbf{f} \cdot \tau \, d\Gamma \qquad \text{for all } \phi \in S_0.$$

If Γ_{03} is empty then the linear functional $G(\cdot)$ takes on the simpler form

$$G(\phi) = -\int_{\Omega} \mathbf{f} \times \operatorname{grad} \phi \, d\Gamma.$$

One may deduce the *essential* and *natural* boundary conditions associated with (11.17)–(11.18) in the usual manner, i.e., through formal integration by parts procedures. These procedures yield that ψ and ω, should these be sufficiently smooth, satisfy the streamfunction-vorticity equations (11.3)–(11.4) and that all boundary conditions involving $\partial\psi/\partial n$ and $\partial\omega/\partial n$ are natural to the weak formulation (11.17)–(11.18). All boundary conditions requiring the specification of ψ or ω are essential and thus have been explicitly imposed on the trial functions. Lastly, the auxiliary conditions (11.13) are also natural to the above weak formulation.

The weak formulation (11.17)–(11.18) is the one in common use in engineering practice and is a special case of a more general formulation discussed in Girault and Raviart [1979 and 1986]. In addition to the ease with which a variety of boundary conditions are satisfied, this formulation also allows for the use of low continuity, i.e., merely continuous, finite element subspaces. This is noteworthy since it is more "natural," from an analysis point of view, to pose a weak problem for the streamfunction-vorticity equations wherein one seeks ψ and ω in subspaces of $W^{2,2}(\Omega) = H^2(\Omega)$ and $L^2(\Omega)$, respectively.

11.4. Finite Element Discretizations

The discretization of (11.17)–(11.18) proceeds in the usual manner except for the treatment of inhomogeneous essential boundary conditions and especially because of the need to determine the constants a_i, $i = 1, ..., m_1$. We assume that Γ is polygonal; finite dimensional subspaces $W_2^h \subset W^{1,2}(\Omega)$, $\tilde{W}_2^h \subset W^{1,2}(\Omega)$, $W_4^h \subset W^{1,4}(\Omega)$, and $\tilde{W}_4^h \subset W^{1,4}(\Omega)$ are chosen that are used to define the sets

$$V^h = \{\zeta \in W_2^h : \zeta = \omega_0^h \text{ on } \Gamma_{02}\}$$

and

$$S^h = \{\phi \in W_4^h : \phi = q_0^h \text{ on } \Gamma_{01} \cup \Gamma_{02}; \quad \phi = q_i^h + a_i^h \text{ on } \Gamma_i, \, i = 1, ..., m,$$
$$a_i^h \in \mathbb{R} \text{ arbitrary except that } a_m^h = 0 \text{ for Case 2}\},$$

and the spaces

$$\tilde{V}_0^h = \{\zeta \in \tilde{W}_2^h : \zeta = 0 \text{ on } \Gamma_{02}\} = V_0 \cap \tilde{W}_2^h.$$

and

$$\tilde{S}_0^h = \{\phi \in \tilde{W}_4^h : \phi = 0 \text{ on } \Gamma_{01} \cup \Gamma_{02}; \quad \phi = a_i^h \text{ on } \Gamma_i, \, i = 1, \dots, m,$$

$$a_i^h \in \mathbb{R} \text{ arbitrary except that } a_m^h = 0 \text{ for Case 2}\} = S_0 \cap \tilde{W}_4^h,$$

where $\omega_0^h \in V^h|_{\Gamma_{02}}$, $q_0^h \in S^h|_{\Gamma_{01} \cup \Gamma_{02}}$, and $q_i^h \in S^h|_{\Gamma_i}$, $i = 1, \dots, m$ are approximations to ω_0, q_0, and q_i, respectively. For example, since ω_0^h need be defined only along Γ_{02}, we may choose it to be the boundary interpolant of ω_0 in V^h restricted to the boundary segment Γ_{02}.

The approximation to the streamfunction and vorticity are then required to satisfy the following problem. We seek $\psi^h \in V^h$ and $\omega^h \in S^h$ such that

$$A(\omega^h, \zeta^h) + B(\zeta^h, \psi^h) = F(\zeta^h) \quad \text{for all } \zeta^h \in \tilde{V}_0^h \quad (11.19)$$

and

$$\nu B(\omega^h, \phi^h) + C(\omega^h, \psi^h, \phi^h) = G(\phi^h) \quad \text{for all } \phi^h \in \tilde{S}_0^h. \quad (11.20)$$

Upon choosing bases for W_2^h, W_4^h, $\tilde{W}_2^h$, and $\tilde{W}_4^h$, the discrete weak formulation (11.19)–(11.20) is equivalent to a system of nonlinear algebraic equations. Let $\{\phi_j\}$, $j = 1, \dots, J$, $\{\zeta_k\}$, $k = 1, \dots, K$, $\{\tilde{\phi}_j\}$, $j = 1, \dots, \tilde{J}$, and $\{\tilde{\zeta}_k\}$, $k = 1, \dots, \tilde{K}$, denote bases for W_2^h, W_4^h, $\tilde{W}_2^h$, and $\tilde{W}_4^h$, respectively. Of course, these sets of basis functions may be all the same if $W_2^h = W_4^h = \tilde{W}_2^h = \tilde{W}_4^h$. Suppose these basis functions are ordered so that

$$\phi_j = 0 \text{ on } \Gamma_{01} \cup \Gamma_{02} \bigcup_{i=1}^{m} \Gamma_i \quad \text{for } j = 1, \dots, N,$$

$$\zeta_k = 0 \text{ on } \Gamma_{02} \quad \text{for } k = 1, \dots, L,$$

$$\tilde{\phi}_j = 0 \text{ on } \Gamma_{01} \cup \Gamma_{02} \bigcup_{i=1}^{m} \Gamma_i \quad \text{for } j = 1, \dots, \tilde{N},$$

and

$$\tilde{\zeta}_k = 0 \text{ on } \Gamma_{02} \quad \text{for } k = 1, \dots, \tilde{L}.$$

Thus these basis functions correspond to all degrees of freedom, both associated with the interior of Ω and the boundary Γ, except for those that correspond to function values on the indicated boundary segments. Furthermore, assume that $\{\phi_j\}$, for $j = N + 1, \dots, N + M_0$, and $\{\tilde{\phi}_j\}$, for $j = \tilde{N} + 1, \dots, \tilde{N} + \tilde{M}_0$, correspond to degrees of freedom associated with function values on $\Gamma_{01} \cup \Gamma_{02}$, and $\{\phi_j\}$, for $j = N + M_{i-1} + 1, \dots, N + M_i$, and $\{\tilde{\phi}_j\}$, for $j = \tilde{N} + \tilde{M}_{i-1} + 1, \dots, \tilde{N} + \tilde{M}_i$, correspond to function value degrees of freedom associated with Γ_i, $i = 1, \dots, m$. Of course, in a practical implementation one would use a more efficient numbering system for the basis functions. With this numbering system, we have that

$$\psi^h = \sum_{j=1}^{N} \psi_j \phi_j(\mathbf{x}) + \sum_{i=0}^{m} \sum_{j=N+1+M_{i-1}}^{N+M_i} q_i(\mathbf{x}_j)\phi_j(\mathbf{x})$$
$$+ \sum_{i=1}^{m_1} a_i^h \left(\sum_{j=N+1+M_{i-1}}^{N+M_i} \phi_j(\mathbf{x}) \right) \tag{11.21}$$

and

$$\omega^h = \sum_{k=1}^{L} \omega_k \zeta_k(\mathbf{x}) + \sum_{k=L+1}^{K} \omega_0(\mathbf{x}_k)\zeta_k(\mathbf{x}), \tag{11.22}$$

where $M_{-1} = 0$ and $\mathbf{x}_j$ denotes the coordinates of the jth node and where we have approximated q_i, $i = 0, ..., m$ and ω_0 by the corresponding boundary interpolants. Also, for the test functions of (11.19)–(11.20) one may choose $\zeta^h = \tilde{\zeta}_k(\mathbf{x})$ for $k = 1 ..., \tilde{L}$, $\phi^h = \tilde{\phi}_j(\mathbf{x})$ for $j = 1, ..., \tilde{N}$, and

$$\phi^h = \sum_{j=\tilde{N}+1+\tilde{M}_{i-1}}^{\tilde{N}+\tilde{M}_i} \tilde{\phi}_j(\mathbf{x}) \quad \text{for } i = 1, ..., m_1, \tag{11.23}$$

where $\tilde{M}_{-1} = 0$. Thus (11.19)–(11.20) represent $(\tilde{N} + \tilde{L} + m_1)$ nonlinear algebraic equations for the unknowns ψ_j, $j = 1, ..., N$; ζ_k, $k = 1, ..., L$; and a_i^h, $i = 1, ..., m_1$. Of course, in general, we would choose the finite element spaces so that $N + L = \tilde{N} + \tilde{L}$.

We will focus on the case $W_2^h = W_4^h = \tilde{W}_2^h = \tilde{W}_4^h$, i.e., the underlying approximating test and trial spaces for the streamfunction and vorticity are the same. In particular, we will consider finite element spaces defined with respect to a triangulation $\mathfrak{I}_h$ of Ω and that consist of piecewise polynomial functions that are continuous over $\bar{\Omega}$. Of special interest will be piecewise linear finite element spaces.

11.5. The Recovery of the Pressure Field

After determining the streamfunction and vorticity fields, it is often of interest to recover from them the primitive variables, i.e., the velocity $\mathbf{u}$ and the pressure p. In Section 13.4 we discuss the issue of primitive variable recovery from a known streamfunction field; much of that discussion applies to the present context as well. Therefore, here we merely outline an algorithm for the recovery of the pressure and velocity from known streamfunction and vorticity fields.

The approximate velocity field is recovered in the obvious manner through the use of the relation $\mathbf{u} = \operatorname{curl} \psi$, i.e., by differentiating, within each element, the computed streamfunction field. The momentum equation (1.1), in the case of plane flow, may be written in the form

$$\operatorname{grad} \mathcal{H} = -\nu \operatorname{curl} \omega + \operatorname{curl} \psi \times \mathbf{k}\omega,$$

where $\mathbf{k}$ is the unit vector perpendicular to the plane of the flow, $\mathcal{H} = p + (\mathbf{u} \cdot \mathbf{u})/2$, and once again the constant density has been absorbed into the pressure p. Clearly, p is known once $\mathcal{H}$ and $\mathbf{u}$ are known; however, it simplifies the calculation to compute $\mathcal{H}$ and $\mathbf{u}$ from ψ and ω instead of computing p directly. For the reasons discussed in Section 13.4, it is necessary to base the calculation of $\mathcal{H}$ on the Stokes problem

$$-\Delta \mathbf{w} + \operatorname{grad} \mathcal{H} = -\nu \operatorname{curl} \omega + \operatorname{curl} \psi \times \mathbf{k}\omega \text{ in } \Omega, \qquad (11.24)$$

$$\operatorname{div} \mathbf{w} = 0 \text{ in } \Omega, \qquad (11.25)$$

and a component of $\mathbf{w}$ vanishes on a portion of the boundary whenever the corresponding component of the velocity $\mathbf{u}$ is specified. For example, $\mathbf{w} = \mathbf{0}$ at portions of the boundary where the velocity is specified.

A discrete field $\mathcal{H}^h$ is found by discretizing (11.24)–(11.25) using any of the stable primitive variable methods of Chapters 1–5. A discrete field $\mathbf{w}^h$, which tends to zero with h, must also be computed. No boundary conditions for $\mathcal{H}$, and thus for p, are needed on boundary segments where the velocity is specified. Of course, the right-hand side of (11.24)–(11.25) is computed from previously determined *approximate* streamfunction and vorticity fields. Again, for more details concerning the discretization of (11.24)–(11.25), consult Section 13.4 as well as Gunzburger and Peterson [1988].

11.6. Available Error Estimates

The error estimates available for finite element approximations to solutions of the streamfunction-vorticity formulation of plane incompressible viscous flows apply only to the case of homogeneous velocity boundary conditions being specified throughout the boundary of Ω, i.e., $\Gamma = \Gamma_0$ and $\mathbf{g}_i = \mathbf{0}$ for $i = 0, \ldots, m$.

If the underlying streamfunction finite element space W_4^h satisfies $W_4^h \subset H^2(\Omega)$, then optimal error estimates are easily derived. However, in this case one can dispense with the vorticity and solve a fourth-order problem involving only the streamfunction. See Chapter 13. Therefore, we turn to discrete spaces that contain functions that are merely continuous across element boundaries.

We assume that the underlying streamfunction and vorticity finite element test and trial spaces are all the same, i.e., $W_2^h = W_4^h = \tilde{W}_2^h = \tilde{W}_4^h = W^h$. Specifically, let $\Omega \subset \mathbb{R}^2$ be a polygonal domain, $\mathfrak{I}_h$ denote a regular triangulation of Ω, and, for a given positive integer k,

$$W^h = \{\phi \in C(\bar{\Omega}) : \phi|_\Delta \in P_k(\Delta), \Delta \in \mathfrak{I}_h\},$$

i.e., we are using piecewise polynomials that are of degree less than or equal to k in each triangle and that are merely continuous across element boundaries. Error estimates for this situation may be found in Girault and Raviart [1986] and are given, for sufficiently smooth ψ and ω, by

$$|\psi - \psi^h|_1 + \|\omega - \omega^h\|_0 \leq Ch^{k-1/2}|\ln h|^\sigma, \tag{11.26}$$

where $\sigma = 1$ for $k = 1$ and $\sigma = 0$ for $k > 1$ and where C is a constant independent of h. This estimate is not optimal, with regard to the power of h, for either the derivatives of the streamfunction, i.e., the velocity components, or for the vorticity. Indeed, (11.26) indicates that one may loose a half power in h for the velocity and three-halves power for the vorticity.

There is some computational and theoretical evidence that the estimate (11.26) is not sharp with regards to the streamfunction error. In Gunzburger and Peterson [1988] are found some computational experiments that indicate that the derivatives of the streamfunctions are better approximated than is indicated by (11.26). Furthermore, in Fix *et al.* [1984] it is shown, for the case of the *linear* Stokes equations, that the derivatives of the streamfunction are essentially optimally approximated in the sense that

$$|\psi - \psi^h|_1 \leq Ch^{k-\varepsilon}, \tag{11.27}$$

where $\varepsilon = 0$ for $k > 1$ and $\varepsilon > 0$ is arbitrary for $k = 1$. However, we note that (11.26) seems to be sharp for the vorticity error and thus vorticity approximations are, in general, very poor. For example, if piecewise linear finite element spaces are used, then the error in the vorticity is $O(h^{1/2})$, while (11.27) indicates that the error in the derivatives of the streamfunction is essentially $O(h)$.

If one is interested in the primitive variables, then the velocity error is given by the error in the derivatives of the streamfunction. Then (11.26) yields

$$\|\mathbf{u} - \mathbf{u}^h\|_0 \leq Ch^{k-1/2}|\ln h|^\sigma,$$

where $\mathbf{u} = \operatorname{curl} \psi$ and $\mathbf{u}^h = \operatorname{curl} \psi^h$, the latter computed element-wise. For the linear case (11.27) gives that

$$\|\mathbf{u} - \mathbf{u}^h\|_0 \leq Ch^{k-\varepsilon}. \tag{11.28}$$

It is likely that (11.28) is also valid for the nonlinear case.

If the pressure is found as indicated in Section 11.5, we then have the estimate (Gunzburger and Peterson [1988])

$$\|p - p^h\|_0 \leq C\bigg(|\psi - \psi^h|_1 + \|\omega - \omega^h\|_0 + |\psi - \psi^h|_1\|\omega - \omega^h\|_0$$

$$+ \inf_{\hat{p}^h \in P^h} \|p - \hat{p}^h\|_0\bigg), \tag{11.29}$$

where C again does not depend on h and where P^h is the finite element space in which one seeks an approximation to $\mathcal{H}$. The last term on the right-hand side of (11.29) arises from the fact that (11.24)–(11.25) is solved discretely and the remaining terms are due to the fact that the right-hand side of (11.24) is itself only known approximately. From (11.29) and the previous discussion of the errors for the streamfunction and vorticity, one sees that the error in the pressure will never be better than the error in the vorticity. Thus, for example, if one uses linear polynomials for the vorticity and streamfunction, then regardless of how one discretizes (11.24)–(11.25), one will not achieve better than $O(h^{1/2})$ accuracy for the pressure. Therefore, in this case, when discretizing (11.24)–(11.25), one should choose among the stable approximating spaces for $\mathbf{w}$ and $\mathcal{H}$ discussed in Chapters 3 and 5, which yield, at best, first-order approximations to $\mathcal{H}$ in the $L^2(\Omega)$-norm.

We close by noting that for various practical as well as theoretical reasons one may not want to choose the degree of the polynomials used for all four test and trial spaces to be the same. The particular choices

$$W_4^h = \tilde{W}_2^h = \{\phi \in C(\bar{\Omega}) : \phi|_\square \in Q_2(\square), \quad \square \in \mathcal{Q}_h\}$$

$$W_2^h = \tilde{W}_4^h = \{\phi \in C(\bar{\Omega}) : \phi|_\square \in Q_1(\square), \quad \square \in \mathcal{Q}_h\} \tag{11.30}$$

for some "triangulation" Q_h of Ω into quadrilaterals is used for some calculations in Habashi *et al.* [1987]; Hafez *et al.* [1987]; and Peeters, Habashi, and Dueck [1987]. Here, the streamfunction trial space and the vorticity test space are based on the use of polynomials of one degree higher than those used for the vorticity trial space and the streamfunction test space. Note that if one chooses $W_2^h = \tilde{W}_2^h \neq W_4^h = \tilde{W}_4^h$ then, in general, $N + L \neq \tilde{N} + \tilde{L}$. Also, although the choice (11.30) has been used in engineering calculations, to this date it has not been subjected to mathematical analyses.

12

Solution Techniques for Multiply Connected Domains

As has been noted earlier, the main difficulty that multiply connected domains present is that the streamfunction can be completely specified on only one portion of the boundary. We have that ψ and ψ^h are known on $\Gamma_{01} \cup \Gamma_{02}$, but are known only up to additive constants on Γ_i, $i = 1, ..., m_1$, where $m_1 = m$ for Case 1 and $m_1 = m - 1$ for Case 2. In the discrete equations (11.19)–(11.20) the constants a_i^h are to be determined as part of the solution procedure. However, solving for these constants simultaneously with the other unknown parameters ψ_j and ω_k appearing in (11.21)–(11.22) requires the use of the basis functions (11.23). The latter are only *semi-local* in the sense that they couple all points on the corresponding boundary segment Γ_i; indeed, in the discrete equations, a_i^h is coupled to all other unknowns associated with elements whose closures intersect with Γ_i. Thus special care must be taken when numbering unknowns so as not to detrimentally affect the structure, e.g., bandwidth, of the discrete system of equations.

A class of alternate solution methods is based on the idea that, instead of using the basis functions (11.23), one guesses the values of the constants a_i^h, $i = 1, ..., m_1$, appearing in (11.19)–(11.20). Then one may solve the latter *mathematically well-posed problem for* ψ^h and ω^h. However, in general (11.13), or rather a discrete approximation of

149

(11.13), will not be satisfied. At this point one may change the guesses for the a_i^h, repeating the above process until convergence is achieved, i.e., until (11.13) is satisfied. Of course, this technique requires the choice of a method to update the a_i^h's and, when coupled with a particular method for solving the nonlinear discrete equations, leads to a variety of composite methods.

Concerning the solution of the discrete nonlinear equations, there are many methods available. If one is interested in preserving the feature of the discretization procedure that led to (11.19)–(11.20), that no artificial boundary conditions on the vorticity be imposed at boundaries at which the velocity is known, then one cannot use algorithms that, e.g., iterate between (11.19) and (11.20). Here we focus on Newton's method for the coupled system (11.19)–(11.20), not necessarily because this is the best method, but because it simplifies the exposition of different treatments of multiply connected domains. As in the primitive variable case, Newton's method is usually used in conjunction with some type of continuation procedure when solutions at high values of the Reynolds number are desired. We refer to the discussion of Chapter 8 for relevant details. We also note, without further elaboration, that, as in the primitive variable case, Newton's method for the discrete streamfunction-vorticity equations converges quadratically for sufficiently accurate initial guesses.

12.1. Newton's Method with the Use of Semi-Local Basis Functions

Newton's method for the solution of (11.19)–(11.20) may be described as follows (Girault and Raviart [1979 and 1986]; Tezduyar, Glowinski, and Liou [1988]; and Gunzburger and Peterson [1988]). Given $\psi^{(0)} \in V^h$ and $\omega^{(0)} \in S^h$, compute the sequence $\{\psi^{(\ell)}, \omega^{(\ell)}\}$ for $\ell \geq 1$ from the sequence of linear algebraic systems: for $\ell = 1, 2, \ldots$,

$$A(\omega^{(\ell)}, \zeta^h) + B(\zeta^h, \psi^{(\ell)}) = F(\zeta^h) \qquad \text{for all } \zeta^h \in \hat{V}_0^h \qquad (12.1)$$

and

$$\nu B(\omega^{(\ell)}, \phi^h) + C(\omega^{(\ell)}, \psi^{(\ell-1)}, \phi^h) + C(\omega^{(\ell-1)}, \psi^{(\ell)}, \phi^h)$$

$$= G(\phi^h) + C(\omega^{(\ell-1)}, \psi^{(\ell-1)}, \phi^h) \qquad \text{for all } \phi^h \in \hat{S}_0^h, \qquad (12.2)$$

where $\psi^{(\ell)}$ and $\omega^{(\ell)}$ are defined by equations such as (11.21) and (11.22) with the unknown coefficients ψ_j, ω_k, and a_i^h now parametrized with ℓ. In (12.2) the set of test functions include the semi-local functions defined in (11.23), and thus a set of constants $\{a_i^h\}$ is calculated at each step of the iteration.

An alternative to Newton's method is the following method, which— although it only converges linearly—often has a larger attraction ball. In fact, as was the case for the primitive variable formulation, for sufficiently small values of the Reynolds number and for simple boundary conditions, this method can be shown to be globally convergent. The sequence $\{\psi^{(\ell)}, \omega^{(\ell)}\}$ for $\ell \geq 1$ is now generated by the equations

$$A(\omega^{(\ell)}, \zeta^h) + B(\zeta^h, \psi^{(\ell)}) = F(\zeta^h) \qquad \text{for all } \zeta^h \in \tilde{V}_0^h \quad (12.3)$$

and

$$\nu B(\omega^{(\ell)}, \phi^h) + C(\omega^{(\ell-1)}, \psi^{(\ell)}, \phi^h) = G(\phi^h) \qquad \text{for all } \phi^h \in \tilde{S}_0^h. \quad (12.4)$$

One may also consider a composite algorithm consisting of first doing a few steps, usually one or two, using (12.3)–(12.4), in order to get into a sufficiently small neighborhood of the solution of (11.19)–(11.20), and then switching to Newton's method. By using a flag variable to zero out the terms in (12.1)–(12.2) that do not appear in (12.3)–(12.4), one may easily implement both of the above methods in the same code. A final note is that the algorithm (12.3)–(12.4), and therefore the composite algorithm as well, does not need the initial guesses $\psi^{(0)}$ and $\omega^{(0)}$ to satisfy any boundary conditions. In particular, for low values of the Reynolds number, one may often choose $\psi^{(0)} = 0$ and $\omega^{(0)} = 0$.

12.2. Solution Methods Using Only Local Basis Functions

If the values of the constants a_i, $i = 1, \ldots, m_1$, are specified, then from a mathematical point of view, e.g., well posedness, there is no difficulty encountered in solving for the streamfunction and vorticity. This observation forms the basis for methods wherein one guesses values for these constants and then solves for a streamfunction-vorticity pair without regard to the auxiliary conditions (11.13). One may subsequently use the latter to obtain an improved guess for the constants.

Of course, even if the constants were known *a priori*, one would still be faced with the task of solving a nonlinear system of discrete equations. Thus, in general, there are two iterative processes needed, one that updates candidate solutions to the discrete system of equations for the streamfunction and vorticity, without any attempt at satisfying (11.13), and the second that updates the values of the constants a_i in order to better comply with (11.13). Depending on what methods are chosen for each of the two iterations and how these methods are combined, many different composite algorithms may be defined.

The particular method about to be described uses Newton's method for finding approximate streamfunctions and vorticities. This will be the "outside" iteration, i.e., each Newton iterate is required to satisfy (11.13), or rather, some numerical quadrature approximation of (11.13), exactly. Full advantage is taken of the facts that each Newton iterate is found by solving a *linear* algebraic system of equations and that (11.13) is *linear* in the vorticity field. Thus the requirement that a Newton iterate satisfy (11.13) may be enforced exactly in a finite number of "inside" iterations to update the a_i's. Of course, implementation of the algorithm does not require the use of the *semi-local* basis functions (11.23), i.e., one maintains, for discretizations of problems posed on multiply connected domains, the same complexity as for those posed on simply connected domain problems.

The first step in describing the algorithm is to redefine the finite element sets that are used. Thus, for a *given* sequence of m_1-tuples $\boldsymbol{\alpha}^{(\ell, s)} = \{\alpha_i^{(\ell, s)}, i = 1, ..., m_1\}$, $s = 0, ..., m_1$ and $\ell = 0, 1, ...$ we now employ the finite element sets

$$V^h = \{\zeta \in W_2^h : \zeta = \omega_0^h \text{ on } \Gamma_{02}\}$$

and

$$S_{\ell, s}^h = \{\phi \in W_4^h : \phi = q_0^h \text{ on } \Gamma_{01} \cup \Gamma_{02}; \phi = q_i^h + \alpha_i^{(\ell, s)} \text{ on } \Gamma_i, i = 1, ..., m,$$

$$\text{where } \alpha_m^{(\ell, s)} = 0 \text{ for Case 2}\},$$

and the spaces

$$\tilde{V}_0^h = \{\zeta \in \tilde{W}_2^h : \zeta = 0 \text{ on } \Gamma_{02}\} = V_0 \cap \tilde{W}_2^h$$

and

$$\tilde{S}_0^h = \{\phi \in \tilde{W}_4^h : \phi = 0 \text{ on } \Gamma_{01} \cup \Gamma_{02} \text{ and on } \Gamma_i, i = 1, ..., m\}.$$

Then, for $\ell = 1, 2, ...$ given the functions $\psi^{(\ell-1)}$ and $\omega^{(\ell-1)}$ and the constants $\alpha^{(\ell, s)} = \{\alpha_i^{(\ell, s)}, i = 1, ..., m_1\}$, $s = 0, ..., m_1$, we define $\psi^{(\ell, s)} \in S_{\ell, s}^h$

and $\omega^{(\ell,s)} \in V^h$, for $s = 0, ..., m$, to be solutions of

$$A(\omega^{(\ell,s)}, \zeta^h) + B(\zeta^h, \psi^{(\ell,s)}) = F(\zeta^h) \quad \text{for all } \zeta^h \in \tilde{V}_0^h \quad (12.5)$$

and

$$\nu B(\omega^{(\ell,s)}, \phi^h) + C(\omega^{(\ell,s)}, \psi^{(\ell-1)}, \phi^h) + C(\omega^{(\ell-1)}, \psi^{(\ell,s)}, \phi^h)$$

$$= G(\phi^h) + C(\omega^{(\ell-1)}, \psi^{(\ell-1)}, \phi^h) \quad \text{for all } \phi^h \in \bar{S}_0^h. \quad (12.6)$$

In general, none of $\omega^{(\ell,s)}$, $s = 0, ..., m_1$, will satisfy even an approximation of (11.13). However, if we let

$$\omega^{(\ell)} = \sum_{s=0}^{m_1} \beta_s^{(\ell)} \omega^{(\ell,s)}, \quad (12.7)$$

where the numbers $\beta_i^{(\ell)}$, $i = 0, ..., m_1$, satisfy

$$\sum_{s=0}^{m_1} \beta_s^{(\ell)} = 1 \quad (12.8)$$

and

$$\sum_{s=0}^{m_1} \beta_s^{(\ell)} \int_{\Gamma_i} \left((\mathbf{g}_i \cdot \mathbf{n}) \omega^{(\ell,s)} - \nu \frac{\partial \omega^{(\ell,s)}}{\partial n} \right) d\Gamma = \int_{\Gamma_i} \mathbf{f} \cdot \tau \, d\Gamma \quad \text{for } i = 1, ..., m_1, \quad (12.9)$$

then, due to the linearity of (11.13) with respect to ω, $\omega^{(\ell)}$ satisfies that set of relations. Note that (12.8)–(12.9) constitute $(m_1 + 1)$ linear algebraic equations for the $(m_1 + 1)$ unknowns $\beta_i^{(\ell)}$, $i = 0, ..., m_1$. Of course, in practice the integrals of (12.9) are approximated through the use of quadrature rules.

Now, let

$$\psi^{(\ell)} = \sum_{s=0}^{m_1} \beta_s^{(\ell)} \psi^{(\ell,s)}. \quad (12.10)$$

With the observation that the boundary conditions for the streamfunction and vorticity are linear in these variables, and that (12.5) and (12.6) are linear in $\psi^{(\ell,s)}$ and $\omega^{(\ell,s)}$, we have that $\psi^{(\ell)}$ and $\omega^{(\ell)}$ defined by (12.10) and (12.7), respectively, satisfy all the aforementioned boundary conditions and the Newton equations

$$A(\omega^{(\ell)}, \zeta^h) + B(\zeta^h, \psi^{(\ell)}) = F(\zeta^h) \quad \text{for all } \zeta^h \in \tilde{V}_0^h$$

and

$$\nu B(\omega^{(\ell)}, \phi^h) + C(\omega^{(\ell)}, \psi^{(\ell-1)}, \phi^h) + C(\omega^{(\ell-1)}, \psi^{(\ell)}, \phi^h)$$

$$= G(\phi^h) + C(\omega^{(\ell-1)}, \psi^{(\ell-1)}, \phi^h) \quad \text{for all } \phi^h \in \bar{S}_0^h.$$

The unknown constants $a_i^{(\ell)}$, $i = 1, \dots, m_1$, are approximated at the ℓth Newton step by

$$a_i^{(\ell)} = \sum_{s=0}^{m_1} \beta_s^{(\ell)} \alpha_i^{(\ell, s)} \quad \text{for } i = 1, \dots, m_1. \tag{12.11}$$

In summary, given the Newton iterate

$$\{ \psi^{(\ell-1)}, \omega^{(\ell-1)}, [a_i^{(\ell-1)}, i = 1, \dots, m_1] \},$$

the subsequent Newton iterate $\{\psi^{(\ell)}, \omega^{(\ell)}, [a_i^{(\ell)}, i = 1, \dots, m_1]\}$ is determined as follows: First solve (12.5)–(12.6) for $\psi^{(\ell, s)}$ and $\omega^{(\ell, s)}$, $s = 0, \dots, m_1$, using arbitrarily selected values for $\alpha_i^{(\ell, s)}$, $i = 1, \dots, m_1$ and $s = 0, \dots, m_1$; then solve (12.8)–(12.9) for the numbers $\beta_i^{(\ell)}$, $i = 0, \dots, m_1$; finally $\psi^{(\ell)}$, $\omega^{(\ell)}$, and $a_i^{(\ell)}$, $i = 1, \dots, m_1$, are computed by (12.10), (12.7), and (12.11), respectively.

Some remarks concerning this algorithm are in order. First note that each Newton step requires the solution of the $m_1 + 1$ linear systems (12.5)–(12.6). However, the left-hand side of these systems are all identical so that the major portion of a direct computation, i.e., the factorization step, need be carried out only once per Newton iteration. The values of the guesses $\alpha_i^{(\ell, s)}$, $i = 1, \dots, m_1$ and $s = 0, \dots, m_1$, needed for each ℓ in order to completely define (12.5)–(12.6) may be arbitrarily chosen, except for the mild requirement that the $(m_1 + 1) \times (m_1 + 1)$ matrix whose sth row is given by $\{1, \alpha_1^{(\ell, s)}, \dots, \alpha_{m_1}^{(\ell, s)}\}$ be nonsingular.

We again emphasize that the algorithm takes explicit advantage of the linearity of the boundary conditions and of the defining systems for the Newton iterates. Also, each Newton iterate is required to satisfy the auxiliary conditions (11.13). The analogous algorithm based on the simpler iterative scheme (12.3)–(12.4) can be easily defined and also combined with the one based on Newton's method to define a hybrid algorithm.

V

THE STREAMFUNCTION FORMULATION

Two-dimensional flows may be described with a model that involves only a single scalar unknown field, compared to two and three for the streamfunction-vorticity and the primitive variable formulations, respectively. The price paid for the reduction to a single unknown is that one must deal with a fourth-order partial differential equation. Such equations are often encountered in structural mechanics problems and thus there has been considerable progress made in the design, analysis, and implementation of finite element algorithms for the approximation of their solutions.

The difficulty that fourth-order problems pose from an algorithmic point of view is that conforming finite element methods require the use of continuously differentiable finite element functions. The study of nonconforming methods has received great impetus from the desire to avoid the construction of finite element spaces, and their bases, that satisfy this requirement. In this part we review both conforming and nonconforming finite element methods for the fourth-order stream-function equations of viscous incompressible flows.

As was the case for the streamfunction-vorticity formulation, we do not consider three-dimensional streamfunction, or vector potential, formulations, except in some remarks in the last chapter.

13

Algorithms for Determining Streamfunction Approximations

One may eliminate the vorticity ω from (11.3)–(11.4) and obtain a formulation for plane, steady, viscous incompressible flow involving only the streamfunction ψ, namely,

$$\nu\Delta^2\psi - \psi_y\,\Delta\psi_x + \psi_x\,\Delta\psi_y = \text{curl}\,\mathbf{f} \text{ in } \Omega. \tag{13.1}$$

We will only consider the no-slip boundary conditions,

$$\psi = \frac{\partial\psi}{\partial n} = 0 \text{ on } \Gamma, \tag{13.2}$$

mainly because the available mathematical results concern themselves with these boundary conditions. For the same reason, we only consider plane flows. Some mathematical aspects of finite element methods for the streamfunction formulation are treated in Girault and Raviart [1979 and 1986]; here, we mostly follow Cayco and Nicolaides [1986 and 1989].

The attractions of the streamfunction formulation are that the incompressibility constraint (1.2) is automatically accounted for and, in $\mathbb{R}^2$, there is only one unknown field to solve for. We will discuss the choice of weak formulation and how it relates to the choice of finite elements, error estimates for some particular finite element spaces, and a method for recovering a pressure approximation once the streamfunction approximation has been determined. With small changes, the discussions of Chapters 11 and 12 concerning alternate boundary conditions and multiply connected domains also apply to the streamfunction formulation.

13.1. Weak Formulations for the Streamfunction Equations

Let

$$H_0^2(\Omega) = \left\{ \psi \in H^2(\Omega) : \psi = \frac{\partial \psi}{\partial n} = 0 \text{ on } \Gamma \right\},$$

$$D_0(\psi, \phi) = \nu \int_\Omega \Delta\psi \, \Delta\phi \, d\Omega$$

and

$$D_1(\psi, \xi, \phi) = \int_\Omega \Delta\psi (\xi_y \phi_x - \xi_x \phi_y) \, d\Omega.$$

Then, a natural weak formulation of (13.1)–(13.2) is to seek $\psi \in H_0^2(\Omega)$ such that

$$D_0(\psi, \phi) + D_1(\psi, \psi, \phi) = (\mathbf{f}, \text{curl } \phi) \quad \text{for all } \phi \in H_0^2(\Omega). \quad (13.3)$$

Conforming finite element approximations are defined in the usual manner, i.e., we choose a finite dimensional subspace $\Psi^h \subset H_0^2(\Omega)$ and then seek a $\psi^h \in \Psi^h$ such that (13.3) holds for all $\phi^h \in \Psi^h$. The inclusion $\Psi^h \subset H_0^2(\Omega)$ requires the use of finite element functions that are continuously differentiable over Ω; this, of course, is a difficult constraint to satisfy in practice.

Because of the stringent $C^1(\Omega)$ requirement of conforming finite element approximations, one is naturally interested in nonconforming approximations for which $\Psi^h \not\subset H_0^2(\Omega)$. Unfortunately, the weak formulation (13.3) does not lend itself to stable nonconforming approximations. The reason for this is that although the bilinear form $D_0(\cdot, \cdot)$ is coercive on $H_0^2(\Omega) \times H_0^2(\Omega)$, i.e., for some constant $\alpha > 0$,

$$D_0(\psi, \psi) \geq \alpha \|\psi\|_2^2 \quad \text{for all } \psi \in H_0^2(\Omega),$$

this form is in general not coercive with respect to nonconforming finite element spaces.

A simple modification of (13.3) results in a useful weak formulation for nonconforming finite element spaces. Let

$$\tilde{D}_0(\psi, \phi) = \nu \int_\Omega (\psi_{xx}\phi_{xx} + 2\psi_{xy}\phi_{xy} + \psi_{yy}\phi_{yy})\, d\Omega$$

and

$$\tilde{D}_1(\psi, \xi, \phi) = \int_\Omega ((\xi_y\psi_{xy} - \xi_x\psi_{yy})\phi_y - (\xi_x\psi_{xy} - \xi_y\psi_{xx})\phi_x)\, d\Omega.$$

We then seek $\psi \in H_0^2(\Omega)$ such that

$$\tilde{D}_0(\psi, \phi) + \tilde{D}_1(\psi, \psi, \phi) = (\mathbf{f}, \operatorname{curl}\phi) \qquad \text{for all } \phi \in H_0^2(\Omega). \quad (13.4)$$

As far as the continuous problem is concerned, it makes no difference whether one uses (13.3) or (13.4); indeed, $\tilde{D}_0(\psi, \phi) = D_0(\psi, \phi)$ and $\tilde{D}_1(\psi, \xi, \phi) = D_1(\psi, \xi, \phi)$ for all $\psi, \phi, \xi \in H_0^2(\Omega)$. However, when using nonconforming approximating subspaces, (13.3) and (13.4) generate different finite element methods, only the second of which is of any use.

13.2. Finite Element Spaces

We now describe some finite element spaces for the streamfunction formulation. The conforming elements can be used with either (13.3) or (13.4); indeed, in this case the two weak formulations yield identical results. The nonconforming elements can be used only with (13.4). As usual, we will consider regular triangulations $\mathfrak{I}_h$ of a polygonal domain Ω. We also recall the notations introduced in Chapter 3. For details concerning these elements and their approximation properties, one may consult Ciarlet [1978] and Lascaux and Lesaint [1975]; with regards to the error estimates, consult Cayco and Nicolaides [1986 and 1989].

Nonconforming Triangular Elements. First consider the *Morley triangular* element where within each triangle $\Delta \in \mathfrak{I}_h$ the functions $\psi^h \in \Psi^h$ are quadratic polynomials. There are six degrees of freedom associated with each triangle, and these are chosen to be the value of ψ^h at the vertices of the triangle and the value of the normal derivative of ψ^h at the midsides. Thus, $\Psi^h|_\Delta = P^2(\Delta)$ and $\dim(\Psi^h|_\Delta) = 6$. See Figures 13.1–13.6 for illustrations of the Morley triangle and the other elements discussed in this section.

Next, we have the two *Fraeijs de Veubeke triangular* elements for which the functions in Ψ^h are cubic polynomials within each triangle.

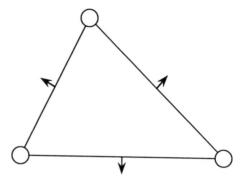

Figure 13.1

The Morley triangular element. ○ Function value degree of freedom, → normal derivative degree of freedom.

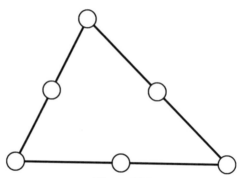

Figure 13.2

The first Fraeijs de Veubeke triangular element. ○ Function value degree of freedom, —— side average of the normal derivative degree of freedom.

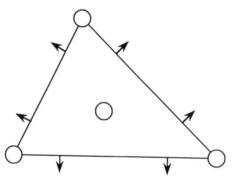

Figure 13.3

The second Fraeijs de Veubeke triangular element. ○ Function value degree of freedom, → normal derivative degree of freedom.

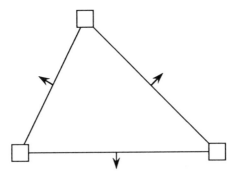

Figure 13.4

The Argyris triangular element. □ Function value, first and second derivatives degrees of freedom, → normal derivative degree of freedom.

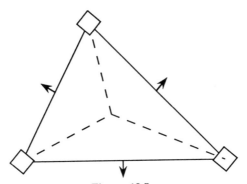

Figure 13.5

The Clough–Tocher triangular element. ◇ Function value and first derivative degrees of freedom, → normal derivative degree of freedom.

Figure 13.6

The Bogner–Fox–Schmit rectangular element. × Function value, first derivative and mixed second derivative degrees of freedom.

For the first of these, there are nine degrees of freedom associated with each triangle; they are chosen to be the function value at the vertices and the midsides of the triangle and the average of the normal derivative on each side. Here, $\dim(\Psi^h|_\Delta) = 9$ and $P^2(\Delta) \subset \Psi^h|_\Delta \subset P^3(\Delta)$. See Figure 13.2.

For the second Fraeijs de Veubeke triangle, there are 10 degrees of freedom associated with each triangle, namely, the function value at the vertices and at the centroid of each triangle and the normal derivative at two points on each side. The two points are chosen to be Gaussian points, i.e., the two points on a side that are at a distance $(3)^{1/2}L/2$ from the midside, where L denotes the length of the side. Here, $\dim(\Psi^h|_\Delta) = 10$ and $\Psi^h|_\Delta = P^3(\Delta)$.

For any of these nonconforming elements, the boundary conditions (13.2) cannot be satisfied exactly, even on polygonal domains. In practice, one usually does the best one can, i.e., one sets to zero all function and normal derivative degrees of freedom associated with the boundary Γ.

Conforming Triangular Elements. It is well known that if one wishes to use a complete piecewise polynomial space with respect to a general triangulation and also to satisfy the inclusion $\Psi^h \subset H^2(\Omega)$, one needs to use piecewise *quintic* polynomials that are *continuously differentiable* over Ω. Such an element is the *Argyris* triangle for which the functions in Ψ^h are quintic polynomials within each triangle $\Delta \in \mathfrak{I}_h$ and the 21 degrees of freedom are chosen to be the function value and the first and second derivatives at the vertices, and the normal derivative at the midsides. Here $\dim(\Psi^h|_\Delta) = 21$ and $\Psi^h|_\Delta = P^5(\Delta)$.

Conforming elements of higher order are also known. Indeed, in Morgan and Scott [1975] and Scott and Vogelius [1985a] are given $C^1(\Omega)$-piecewise polynomial elements of degree greater than or equal to five, defined on arbitrary quasi-uniform triangulations of a polygonal domain Ω. Moreover, the construction of local basis functions is also discussed.

It is possible to achieve the $C^1(\Omega)$ continuity required for the inclusion $\Psi^h \subset H^2(\Omega)$ to hold and still use cubic polynomials; however, to do so, one needs to introduce *macro-elements*. An example is provided by the *Clough–Tocher* triangle. Here we subdivide each triangle $\Delta \in \mathfrak{I}_h$ into three triangles by joining the vertices to the centroid. In each of the smaller triangles, the functions belonging to Ψ^h are cubic polynomials.

There are then 30 degrees of freedom needed to determine the three different cubic polynomials associated with the three triangles within Δ. Eighteen of these are used to ensure that, within Δ, the functions belonging to Ψ^h are continuously differentiable. The remaining 12 degrees of freedom are chosen to be the function value and the first derivatives at the vertices and the normal derivative at the mid-sides. With these degrees of freedom we have that $\Psi^h \subset H^2(\Omega)$; also, $\dim(\Psi^h|_\Delta) = 12$ and $P^3(\Delta) \subset \Psi^h|_\Delta$.

For all the conforming elements, in order to satisfy the stronger inclusion $\Psi^h \subset H_0^2(\Omega)$, one merely sets the function values equal to zero at all vertices on the boundary Γ and the normal derivative equal to zero at all midside nodes and vertices on the boundary Γ.

Conforming Bicubic Rectangles. Suppose now that Ω is such that it may be subdivided into a "triangulation" $\mathcal{Q}_h$ consisting of rectangles. For the *Bogner-Fox-Schmidt* rectangular element we have that the functions belonging to Ψ^h are bicubic polynomials within each $\square \in \mathcal{Q}_h$. The degrees of freedom are chosen to be the function value, the first derivatives, and the mixed second derivative at the vertices. This element satisfies $\Psi^h \subset H^2(\Omega)$, $\dim(\Psi^h|_\square) = 16$, and $\Psi^h|_\square = Q^3(\square)$. Again, for $\Psi^h \subset H_0^2(\Omega)$, we merely set the function and the normal derivative values equal to zero at all vertices on the boundary Γ.

Another conforming bicubic element is defined as the tensor product of cubic splines. These functions are bicubic polynomials within each rectangle, are twice continuously differentiable over $\bar{\Omega}$, and their degrees of freedom are the function values at the nodes (plus some additional ones on the boundary). The support of the most local basis functions for the space of cubic splines (the B-spline functions) is greater than that for the Hermite type Bogner-Fox–Schmit element. However, the simplicity of the degrees of freedom makes this element a popular choice. See Schultz [1973] for a thorough discussion of cubic splines.

13.3. Error Estimates

For a set $K \subset \mathbb{R}^2$ denote by $\|\cdot\|_{0,K}$ the $L^2(K)$ norm, i.e.,

$$\|\phi\|_{0,K} = \left(\int_K \phi^2 \, d\Omega \right)^{1/2}$$

and define the semi-norms

$$|\phi|_{m,K} = (\sum \|D^m\phi\|_{0,K}^2)^{1/2},$$

where the sum extends over all derivatives of order m. Then we may define the semi-norms

$$|\phi|_{m,h} = \left(\sum_{\Delta \in \mathfrak{I}_h} |\phi|_{m,\Delta}^2\right)^{1/2}.$$

Note that if $\phi \in H^m(\Omega)$, then $|\phi|_{m,h} = |\phi|_m$, where

$$|\phi|_m = (\sum \|D^m\phi\|_0^2)^{1/2},$$

where the sum extends over all derivatives of order m.

The available error estimates are of the form

$$|\psi - \psi^h|_{m,h} = O(h^r) \quad \text{provided } \psi \in H^s(\Omega) \cup H_0^2(\Omega), \quad (13.5)$$

where $m = 0$, 1, or 2, and where ψ and ψ^h denote the exact solution of (13.1)–(13.2) and the finite element approximation, respectively. For each of the elements described above, we give, in Table 13.1, the obtainable values of r in the error estimate (13.5), and the value of s required for the error estimate with $m = 2$ to hold. For the error estimates with $m = 0$ or 1, a larger value of s, i.e., more smoothness, may be required.

In practical calculations we have found the Clough–Tocher element to be useful, although it should be noted that no explicit basis is known for that element, i.e., to evaluate a basis function at any point in Ω one must solve a matrix problem.

Table 13.1. Accuracy of finite elements for the streamfunction formulation.

Element	r for $m = 2$	r for $m = 1$	r for $m = 0$	s for $m = 2$
Morley triangle	1	2	2	3
Fraeijs de Veubeke I	1	2	2	3
Fraeijs de Veubeke II	1	2	2	3
Argyris triangle	4	5	6	6
Clough–Tocher triangle	2	3	4	4
Bogner–Fox–Schmit rectangle	2	3	4	4
Bicubic spline rectangle	2	3	4	4

13.4. Recovery of the Velocity and Pressure

Once an approximate streamfunction ψ^h has been determined, one may define an approximate velocity by $\mathbf{u}^h = \operatorname{curl} \psi^h$, i.e., $\mathbf{u}^h = (\partial \psi^h / \partial y, -\partial \psi^h / \partial x)$, where the differentiations are carried out within an element. Since, of course, $\mathbf{u} = \operatorname{curl} \psi$, an estimate for the $L^2(\Omega)$-norm of the error in the approximate velocity is provided by the error estimate (13.5), with $m = 1$, for the streamfunction.

One may also be interested in recovering an approximation for the pressure. The most common technique is to take the divergence of the momentum equation (1.1), resulting in a Poisson equation for the pressure wherein the data depends on the velocity field and its derivatives. The latter may be determined from the streamfunction as indicated above. Specifically, we then have (if $\mathbf{f}$ vanishes)

$$\Delta p = G(\psi) = -2\left(\frac{\partial^2 \psi}{\partial x_1 \, \partial x_2}\right)^2 + 2\left(\frac{\partial^2 \psi}{\partial x_1^2}\right)\left(\frac{\partial^2 \psi}{\partial x_2^2}\right). \qquad (13.6)$$

Of course, one then needs to define a boundary condition for the pressure. The most rational choice (Gresho and Sani [1988]) is to take the inner product of the momentum equation with the unit outer normal vector to Γ, thus generating a boundary condition for $\partial p/\partial n$, where again the data depends on the velocity and its derivatives. Such a procedure can be implemented in a finite element context based on the weak formulation wherein one seeks a $p \in H^1(\Omega)$ such that

$$\int_\Omega \operatorname{grad} p \cdot \operatorname{grad} q \, d\Omega = -\int_\Omega G(\psi) q \, d\Omega \qquad \text{for all } q \in H^1(\Omega). \quad (13.7)$$

Moreover, the Neumann boundary condition for the pressure discussed above is natural to the weak formulation (13.7).

At first glance one might question the validity of applying the normal momentum equation at the boundary in order to define a boundary condition, although it may be argued that this step is justified, at least whenever $\operatorname{div} \mathbf{u} = 0$ in Ω and $\mathbf{u} \cdot \mathbf{n} = 0$ on Γ. See, e.g., Gresho and Sani [1988]. Note that solving a Poisson equation leads one to seek an approximate pressure in subspaces of $H^1(\Omega) \cap L_0^2(\Omega)$. On the other hand, having sought a streamfunction belonging to $H_0^2(\Omega)$ indicates that the velocity should belong to $\mathbf{H}_0^1(\Omega)$ and thus it seems more natural to seek a pressure in $L_0^2(\Omega)$. A serious objection to determining the pressure from (13.6) is that the data for this Poisson equation is obtained by differentiating the momentum equation and involves second derivatives of the streamfunction. Therefore this data is not always well defined for some choices of approximating finite element subspaces for the streamfunction.

There is an alternative to the above procedure that suffers from none of these problems. This procedure is based on the weak form (1.9) of the momentum equation. We describe it here for the case of *conforming* streamfunction approximations; for the nonconforming case, see Cayco [1985].

Suppose the velocity field $\mathbf{u} = \operatorname{curl} \psi$ is known; then (1.9) may be viewed as an equation determining the pressure. i.e., we seek a $p \in L_0^2(\Omega)$ such that

$$b(\mathbf{v}, p) = g(\mathbf{f}, \psi; \mathbf{v}) \quad \text{for all } \mathbf{v} \in \mathbf{H}_0^1(\Omega), \tag{13.8}$$

where the linear functional $g(\mathbf{f}, \psi; \cdot)$ on $\mathbf{H}_0^1(\Omega)$ is given by

$$g(\mathbf{f}, \psi; \mathbf{v}) = (\mathbf{f}, \mathbf{v}) - a(\operatorname{curl} \psi, \mathbf{v}) - c(\operatorname{curl} \psi, \operatorname{curl} \psi, \mathbf{v}). \tag{13.9}$$

In (13.8)–(13.9), the forms $a(\cdot, \cdot)$, $b(\cdot, \cdot)$, and $c(\cdot, \cdot, \cdot)$ are defined as in, e.g., (1.6)–(1.8). It can be shown that, given $\psi \in H_0^2(\Omega)$ and $\mathbf{f} \in \mathbf{L}^2(\Omega)$, $p \in L_0^2(\Omega)$ is uniquely determined by (13.8). Note that (13.8) determines the pressure without the need of devising any boundary conditions for the pressure, without differentiating the momentum equation, and without imposing the normal momentum equation on the boundary.

Having obtained a discrete streamfunction ψ^h, one may discretize (13.8) in the usual manner in order to obtain an approximation for the pressure. Specifically, one chooses finite element subspaces $S_0^h \subset L_0^2(\Omega)$ and $\mathbf{V}_0^h \subset \mathbf{H}_0^1(\Omega)$ and then seeks a $p^h \in S_0^h$ such that

$$b(\mathbf{v}^h, p^h) = g(\mathbf{f}, \psi^h; \mathbf{v}^h) \quad \text{for all } \mathbf{v}^h \in \mathbf{V}_0^h. \tag{13.10}$$

At this point it seems that in order to have a well-defined and stable approximate pressure, one merely needs to choose the trial space S_0^h for the pressure and the test space $\mathbf{V}_0^h$ in a manner such that the div-stability condition (2.10) is satisfied. However, the fact that in practice we only have available an approximation ψ^h to ψ poses a problem. Note that for those $\mathbf{v}^h \in \mathbf{V}_0^h$ that are discretely divergence free, i.e., for $\mathbf{v}^h \in \mathbf{Z}^h$, the left-hand side of (13.10) vanishes, but in general, the right-hand side $g(\mathbf{f}, \psi^h; \mathbf{v}^h)$ does not, i.e., the right-hand side of (13.10) is not in the range of the discrete gradient operator defined by the left-hand side of that equation.

We therefore consider, instead of (13.8), the following equivalent Stokes problem: Given $\mathbf{f} \in \mathbf{L}^2(\Omega)$ and $\psi \in H_0^2(\Omega)$, seek $\mathbf{w} \in H_0^1(\Omega)$ and $p \in L_0^2(\Omega)$ such that

$$a(\mathbf{w}, \mathbf{v}) + b(\mathbf{v}, p) = g(\mathbf{f}, \psi; \mathbf{v}) \qquad \text{for all } \mathbf{v} \in \mathbf{H}_0^1(\Omega) \qquad (13.11)$$

and

$$b(\mathbf{w}, q) = 0 \qquad \text{for all } q \in L_0^2(\Omega), \qquad (13.12)$$

where $g(\mathbf{f}, \psi; \mathbf{v})$ is still given by (13.9). This problem is uniquely solvable; moreover, $\mathbf{w} = \mathbf{0}$ and p coincides with the solution of (13.8). Of course, the auxiliary variable $\mathbf{w}$ is not the velocity $\mathbf{u} = \text{curl } \psi$.

By choosing finite element subspaces $S_0^h \subset L_0^2(\Omega)$ and $\mathbf{V}_0^h \subset \mathbf{H}_0^1(\Omega)$, we may define a discrete version of (13.11)–(13.12): Given $\psi^h \in \Psi^h \subset H_0^2(\Omega)$ and $\mathbf{f} \in \mathbf{L}^2(\Omega)$, seek $\mathbf{w}^h \in \mathbf{V}_0^h$ and $p^h \in S_0^h$ such that

$$a(\mathbf{w}^h, \mathbf{v}^h) + b(\mathbf{v}^h, p^h) = g(\mathbf{f}, \psi^h; \mathbf{v}^h) \qquad \text{for all } \mathbf{v}^h \in \mathbf{V}_0^h \qquad (13.13)$$

and

$$b(\mathbf{w}^h, q^h) = 0 \qquad \text{for all } q^h \in S_0^h. \qquad (13.14)$$

If the approximating spaces $\mathbf{V}_0^h$ and S_0^h satisfy the div-stability condition (2.10), then (13.13)–(13.14) have a unique solution; moreover, we have the error estimate

$$|\mathbf{w}^h|_1 + \|p - p^h\|_0 \le C_1 \inf_{q^h \in S_0^h} (\|p - q^h\|_0) + C_2 |\psi - \psi^h|_2. \quad (13.15)$$

Thus we see that as $h \to 0$, $\mathbf{w}^h \to \mathbf{0}$, and $p^h \to p$.

The appropriate choice of pressure space S_0^h depends on the choice of streamfunction space Ψ^h in the sense that it is efficient to equilibrate the rates of convergence of the two terms on the right-hand side of (13.15). Of course, the pressure space must also be chosen in conjunction with a velocity space so that the div-stability condition is satisfied. Thus, for example, if the streamfunction is approximated using the Clough–Tocher triangle, the Taylor–Hood element could be used for $\mathbf{w}^h$ and p^h. In this case, both terms on the right-hand side of (13.15) are $O(h^2)$, and p^h is a second-order approximation, in the $L^2(\Omega)$-norm, to the pressure p. See Cayco [1985] for details and also for proper pressure-velocity spaces accompanying the nonconforming streamfunction spaces discussed in Section 13.2.

VI

EIGENVALUE PROBLEMS CONNECTED WITH STABILITY STUDIES FOR VISCOUS FLOWS

We now consider the application of some of the techniques and theories connected with finite element methods for the approximate solution of the Navier–Stokes equations to eigenvalue problems associated with the study of the stability of viscous flows. We will focus on the primitive variable formulation; analogous results may be obtained for the streamfunction-vorticity and streamfunction settings as well.

14

Energy Stability Analysis of Viscous Flows

14.1. The Eigenvalue Problem Associated with Energy Stability

Consider a given flow $\mathbf{u}$ in a region Ω that at some instant in time, denoted by $t = 0$, has a velocity field $\mathbf{u}(\mathbf{x}, 0) = \mathbf{U}(\mathbf{x})$. Suppose that at this instant in time the given flow is subject to a perturbation $\mathbf{w}(\mathbf{x}, 0)$. The subsequent departure of the perturbed flow from the given flow is denoted by $\mathbf{w}(\mathbf{x}, t)$ so that the perturbed flow is given by $\mathbf{u}(\mathbf{x}, t) + \mathbf{w}(\mathbf{x}, t)$, where $\mathbf{u}$ denotes the subsequent unperturbed flow. Nothing is assumed concerning the size of the initial perturbation $\mathbf{w}(\mathbf{x}, 0)$ relative to the size of the given flow $\mathbf{U}(\mathbf{x})$. We assume that both the unperturbed flow $\mathbf{u}$ and the perturbed flow $\mathbf{u} + \mathbf{w}$ satisfy the unsteady Navier–Stokes equations, the continuity equation, and have the same, possibly inhomogeneous, values on the boundary Γ. Thus the perturbation $\mathbf{w}$ satisfies the incompressibility constraint and vanishes on Γ; however, due to the nonlinear convection term, $\mathbf{w}$ does not satisfy the unsteady Navier–Stokes equations.

The kinetic energy $\mathcal{K}(t)$ of the perturbation $\mathbf{w}(\mathbf{x}, t)$ is defined by

$$\mathcal{K}(t) = \frac{1}{2} \int_\Omega \mathbf{w} \cdot \mathbf{w} \, d\Omega,$$

where we assume that all variables have been nondimensionalized so that the constant density may be set to unity. We say that the given flow **U** is *stable in the energy sense* if $\mathcal{K}(t)$ tends to zero as t increases. Using the facts that both the unperturbed flow **u** and the perturbed flow **u** + **w** satisfy the continuity condition, the Navier–Stokes equation, and the same boundary condition, we are led to

$$\frac{d\mathcal{K}}{dt} = -\int_{\Omega} [\mathbf{w} \cdot D(\mathbf{U}) \cdot \mathbf{w} + \nu \operatorname{grad} \mathbf{w} : \operatorname{grad} \mathbf{w}] \, d\Omega, \qquad (14.1)$$

where

$$D(\mathbf{U}) = \tfrac{1}{2}(\operatorname{grad} \mathbf{U} + (\operatorname{grad} \mathbf{U})^T)$$

is the rate of strain tensor of the given flow **U**. Since we are assuming that all variables have been nondimensionalized, ν in (14.1) is the inverse of the Reynolds number $Re = \rho VL/\mu$, where μ is the viscosity coefficient, ρ the constant density, L a length scale for the problem, e.g., the diameter of Ω, and V a velocity scale, e.g., the maximum value, over Ω, of $\mathbf{U}(\mathbf{x})$.

If the right-hand side of (14.1) is negative, then $\mathcal{K}$ will decrease as t increases. Now, let

$$\tilde{\nu} = \max_{\mathbf{v}}\left(-\frac{\int_{\Omega} \mathbf{v} \cdot D(\mathbf{U}) \cdot \mathbf{v} \, d\Omega}{\int_{\Omega} \operatorname{grad} \mathbf{v} : \operatorname{grad} \mathbf{v} \, d\Omega}\right), \qquad (14.2)$$

where the maximum is sought over all vector fields **v** satisfying $\operatorname{div} \mathbf{v} = 0$ in Ω and $\mathbf{v} = \mathbf{0}$ on Γ, the boundary of Ω. The allowable perturbation **w** satisfies these two constraints so that (14.2) implies that

$$-\int_{\Omega} \mathbf{w} \cdot D(\mathbf{U}) \cdot \mathbf{w} \, d\Omega \le \tilde{\nu} \int_{\Omega} \operatorname{grad} \mathbf{w} : \operatorname{grad} \mathbf{w} \, d\Omega.$$

Combining with (14.1) yields

$$\frac{d\mathcal{K}}{dt} \le -(\nu - \tilde{\nu}) \int_{\Omega} \operatorname{grad} \mathbf{w} : \operatorname{grad} \mathbf{w} \, d\Omega \qquad (14.3)$$

so that if the solution $\tilde{\nu}$ of the maximization problem (14.2) satisfies $\tilde{\nu} < \nu = 1/Re$, where Re is the Reynolds number of the given flow $\mathbf{U}(\mathbf{x})$, then $d\mathcal{K}/dt < 0$ and the flow is stable.

We can actually obtain more information in the nature of an

estimate for the decay rate for $\mathcal{K}$. Let

$$\tilde{\alpha} = \min_{y} \frac{\int_{\Omega} \operatorname{grad} \mathbf{y} : \operatorname{grad} \mathbf{y} \, d\Omega}{\int_{\Omega} \mathbf{y} \cdot \mathbf{y} \, d\Omega}, \qquad (14.4)$$

where the minimum is sought over all vector fields $\mathbf{y}$ satisfying $\operatorname{div} \mathbf{y} = 0$ in Ω and $\mathbf{y} = \mathbf{0}$ on Γ. Then, for all allowable perturbations $\mathbf{w}$, we have that

$$\int_{\Omega} \operatorname{grad} \mathbf{w} : \operatorname{grad} \mathbf{w} \, d\Omega \geq \tilde{\alpha} \int_{\Omega} \mathbf{w} \cdot \mathbf{w} \, d\Omega = 2\tilde{\alpha}\mathcal{K}. \qquad (14.5)$$

Of course, (14.5) is simply the Poincaré inequality (Babuska and Aziz [1972], Ciarlet [1978], or Girault and Raviart [1986]) with the solution of (14.4) providing the best constant. Combining with (14.3) we have that, if $\tilde{\nu} < \nu$,

$$\frac{d\mathcal{K}}{dt} \leq -2\tilde{\alpha}(\nu - \tilde{\nu})\mathcal{K}$$

so that

$$\mathcal{K}(t) \leq \mathcal{K}(0) \exp\{-2\tilde{\alpha}(\nu - \tilde{\nu})t\},$$

i.e., we get exponential decay of the kinetic energy. For details, see Drazin and Reid [1981], Georgescu [1985], Joseph [1976], and Serrin [1958 and 1959].

Summarizing, to see if $d\mathcal{K}/dt < 0$, we need to solve (14.2). If in addition we want to get an estimate for the decay rate, we need also solve (14.4).

The Euler equations corresponding to the maximization problem (14.2) are given by

$$\lambda \, \Delta\mathbf{w} - \operatorname{grad} s = D(\mathbf{U}) \cdot \mathbf{w} \text{ in } \Omega, \qquad (14.6)$$

$$\operatorname{div} \mathbf{w} = 0 \text{ in } \Omega, \qquad (14.7)$$

and

$$\mathbf{w} = \mathbf{0} \text{ on } \Gamma, \qquad (14.8)$$

where $s(\mathbf{x})$ is a Lagrange multiplier associated with the constraint $\operatorname{div} \mathbf{w} = 0$. Given a velocity field $\mathbf{U}(\mathbf{x})$, (14.6)–(14.8) is a *self-adjoint linear eigenvalue problem* for the triple $\mathbf{w}(\mathbf{x}) \neq \mathbf{0}$, $s(\mathbf{x}) \neq 0$, and $\lambda \in \mathbb{R}$.

The solution $\tilde{v}$ of the maximization problem (14.2) is then given by the largest eigenvalue of (14.6)–(14.8). The existence of a non-negative $\tilde{v}$ follows from the fact that the $\Im_{tace}[D(\mathbf{U})] = \text{div}(\mathbf{U}) = 0$. Then *if $v > \tilde{v}$, the given flow $\mathbf{U}$ is stable.* As it is with most stability criteria, no information concerning the stability of the given flow is conveyed when $v \le \tilde{v}$.

We emphasize that the eigenvalue problem (14.6)–(14.8) is *linear* even though no assumption was made concerning the size of the perturbation; in particular, one need not be restricted, as in linearized stability theories, to infinitesimal perturbations. Of course, the linearity of the problem (14.6)–(14.8) is a consequence of the quadratic nature of the nonlinearity of the Navier–Stokes equations. For details concerning the derivation of (14.6)–(14.8), see the references cited above.

Even for simple flows $\mathbf{U}$ in simple regions Ω, it is not possible to determine the eigenvalues of (14.6)–(14.8), except through numerical procedures. Finite element methods for the approximation of the eigenvalues of (14.6)–(14.8) were considered in Peterson [1980 and 1983]. In order to define such approximations, one first recasts (14.6)–(14.8) into the following weak form: Given $\mathbf{U} \in \mathbf{H}'(\Omega)$ for some positive integer r, seek $\mathbf{0} \ne \mathbf{w} \in \mathbf{H}_0^1(\Omega)$, $0 \ne s \in L_0^2(\Omega)$, and $\lambda \in \mathbb{R}$ such that

$$\lambda \tilde{a}(\mathbf{w}, \mathbf{v}) + b(\mathbf{v}, s) = d(\mathbf{U}; \mathbf{w}, \mathbf{v}) \qquad \text{for all } \mathbf{v} \in \mathbf{H}_0^1(\Omega) \qquad (14.9)$$

and

$$b(\mathbf{w}, q) = 0 \qquad \text{for all } q \in L_0^2(\Omega), \qquad (14.10)$$

where $\tilde{a}(\cdot, \cdot) = a(\cdot, \cdot)/v$ and the bilinear forms $a(\cdot, \cdot)$ and $b(\cdot, \cdot)$ are defined in (1.6) and (1.7), respectively, and where

$$d(\mathbf{U}; \mathbf{w}, \mathbf{v}) = -\int_\Omega \mathbf{w} \cdot D(\mathbf{U}) \cdot \mathbf{v} \, d\Omega.$$

Thus we are interested in finding an approximation for $\tilde{v}$, which now denotes the largest eigenvalue of (14.9)–(14.10). We denote by m the algebraic multiplicity of the eigenvalue $\tilde{v}$ and by $\mathscr{E}(\tilde{v})$ the space spanned by the eigenvectors $(\mathbf{w}, s)$ of (14.9)–(14.10) corresponding to the eigenvalue $\tilde{v}$. Due to the fact that (14.9)–(14.10) is self-adjoint, m is also the geometric multiplicity of the eigenvalue $\tilde{v}$ and the dimension of the eigenspace $\mathscr{E}(\tilde{v})$. It can also be shown by standard techniques of elliptic regularity theory that whenever the solution $\mathbf{U}$ of the Navier–Stokes equations belongs to $\mathbf{H}'(\Omega)$, then the eigenfunctions of (14.9)–(14.10) satisfy $\mathbf{w} \in \mathbf{H}^{r+1}(\Omega) \cap \mathbf{H}_0^1(\Omega)$ and $s \in H'(\Omega) \cap L_0^2(\Omega)$.

14.2. Finite Element Approximations

In order to find an approximation for $\tilde{v}$, we consider the following problem: Given $\mathbf{U} \in \mathbf{H}'(\Omega)$, seek $\mathbf{0} \neq \mathbf{w}^h \in \mathbf{V}_0^h \subset \mathbf{H}_0^1(\Omega)$, $0 \neq s^h \in S_0^h \subset L_0^2(\Omega)$, and $\lambda^h \in \mathbb{R}$ such that

$$\lambda^h \tilde{a}(\mathbf{w}^h, \mathbf{v}^h) + b(\mathbf{v}^h, s^h) = d(\mathbf{U}, \mathbf{w}^h, \mathbf{v}^h) \quad \text{for all } \mathbf{v}^h \in \mathbf{V}_0^h \quad (14.11)$$

and

$$b(\mathbf{w}^h, q^h) = 0 \quad \text{for all } q^h \in S_0^h. \quad (14.12)$$

For details concerning the derivation of (14.9)–(14.10) and (14.11)–(14.12), see Peterson [1980 and 1983]. The problem (14.11)–(14.12) falls within the scope of mixed finite element methods for eigenvalue problems. This class of problems has been analyzed in Mercier *et al.* [1981], Osborn [1979], and Peterson [1980]; the application of these analyses to (14.11)–(14.12) can be found in Peterson [1980 and 1983], where, in particular, error estimates for approximations to the eigenvalue $\tilde{v}$ are provided.

We assume that the bilinear forms $a(\cdot, \cdot)$ and $b(\cdot, \cdot)$ and the approximating subspaces $\mathbf{V}_0^h$ and S_0^h satisfy all the conditions that were required of them in Chapter 2 where approximations of the stationary Navier–Stokes equations were considered. In particular, the subspaces $\mathbf{V}_0^h$ and S_0^h are assumed to satisfy the div-stability condition (2.10). We also assume that for a given velocity field $\mathbf{U}$, the bilinear form $d(\mathbf{U}; \mathbf{w}, \mathbf{v})$ is continuous for all $\mathbf{v}, \mathbf{w} \in \mathbf{H}_0^1(\Omega)$; this assumption is valid whenever, e.g., $\mathbf{U} \in \mathbf{H}^1(\Omega)$. Then of interest here are the following results of Peterson [1980 and 1983]. First, there are exactly m eigenvalues of (14.11)–(14.12), counted according to multiplicity, which as the discretization parameter h tends to zero, converge to the eigenvalue $\tilde{v}$ of (14.9)–(14.10). Denote these m eigenvalues by $\tilde{v}_j^h$, $j = 1, \ldots, m$. In addition, we also have the error estimate: For h sufficiently small, there exists a constant C such that

$$|\tilde{v} - \tilde{v}_j^h| \leq C(\varepsilon^h)^2 \quad \text{for } j = 1, \ldots, m, \quad (14.13)$$

where

$$\varepsilon^h = \sup_{\substack{(\mathbf{w}, s) \in \mathcal{E}(\tilde{v}) \\ |\mathbf{w}|_1 + \|s\|_0 = 1}} \inf_{\substack{\mathbf{v}^h \in \mathbf{V}_0^h \\ q^h \in S_0^h}} (|\mathbf{w} - \mathbf{v}^h|_1 + \|s - q^h\|_0). \quad (14.14)$$

Once the regularity of the eigenfunctions belonging to $(\tilde{v})$ has been determined, the right-hand side of (14.14) is purely approximation theoretic. Some concrete examples are provided as follows. Suppose first that $\mathbf{V}_0^h$ and S_0^h are any of the pair of finite element spaces discussed in Section 3.1, i.e., having stable piecewise linear or bilinear velocity fields. Then,

$$|\tilde{v} - \tilde{v}_j^h| = O(h^2) \quad \text{for } j = 1, ..., m, \tag{14.15}$$

whenever $\mathbf{w} \in \mathbf{H}^2(\Omega) \cap \mathbf{H}_0^1(\Omega)$ and $s \in H^1(\Omega) \cap L_0^2(\Omega)$. On the other hand, the Taylor–Hood element of Section 3.2 yields

$$|\tilde{v} - \tilde{v}_j^h| = O(h^4) \quad \text{for } j = 1, ..., m, \tag{14.16}$$

whenever $\mathbf{w} \in \mathbf{H}^3(\Omega) \cap \mathbf{H}_0^1(\Omega)$ and $s \in H^2(\Omega) \cap L_0^2(\Omega)$.

From (14.13)–(14.16), we see that the usual situation concerning eigenvalue approximations by finite element methods is obtained in the present case, namely, that the error in the eigenvalue is the square of the error for the eigenfunction, the latter being measured in the "natural" norm of the problem.

If one is interested in estimating the decay rate as well, one may repeat the above procedure for the Euler equations corresponding to (14.4), namely,

$$\Delta \mathbf{y} - \operatorname{grad} r + \sigma \mathbf{y} = 0 \text{ in } \Omega, \tag{14.17}$$

$$\operatorname{div} \mathbf{y} = 0 \text{ in } \Omega \tag{14.18}$$

and

$$\mathbf{y} = \mathbf{0} \text{ on } \Gamma, \tag{14.19}$$

where again $r(\mathbf{x})$ is the Lagrange multiplier corresponding to the constraint $\operatorname{div} \mathbf{y} = 0$. Also, (14.17)–(14.19) is a linear self-adjoint eigenvalue problem for the triple $\mathbf{y}(\mathbf{x}) \neq \mathbf{0}$, $r(\mathbf{x}) \neq 0$, and $\sigma \in \mathbb{R}$. The best Poincaré constant $\tilde{\alpha}$ is then the minimum eigenvalue of (14.17)–(14.19). A weak formulation and discretization of (14.17)–(14.19) may be defined in a manner entirely analogous to those given above for (14.6)–(14.8). Also, error estimates analogous to (14.13), (14.15), and (14.16) hold for the discrete approximation to $\tilde{\alpha}$.

Solution algorithms for solving the discrete generalized eigenvalue problem (14.11)–(14.12) are discussed in Peterson [1980 and 1983].

15

Linearized Stability Analysis of Stationary Viscous Flows

15.1. The Eigenvalue Problem Associated with Linearized Stability

The object is to study the stability of a given *stationary* flow field $\mathbf{U}(\mathbf{x})$ in the presence of *infinitesimal* perturbations. The perturbed flow is assumed to satisfy the unsteady Navier–Stokes equations and the continuity equation and to take on the same boundary values as $\mathbf{U}(\mathbf{x})$. For the linearized stability analysis, one further assumes that the perturbed flow has the form

$$\mathbf{U}(\mathbf{x}) + \mathbf{w}(\mathbf{x})e^{-\lambda t} \tag{15.1}$$

and, of course, that the perturbation $\mathbf{w}$ is "small" compared to the given flow $\mathbf{U}$. Similar assumptions are made concerning the pressure P and its perturbation s, i.e., the perturbed pressure is given by

$$P(\mathbf{x}) + s(\mathbf{x})e^{-\lambda t}. \tag{15.2}$$

Substitute (15.1)–(15.2) into the equations for unsteady viscous flows and subsequently neglect all quadratic terms in $\mathbf{w}$ and s or their derivatives to obtain the *non-self-adjoint* eigenvalue problem

$$-\nu \, \Delta \mathbf{w} + \mathbf{U} \cdot \operatorname{grad} \mathbf{w} + \mathbf{w} \cdot \operatorname{grad} \mathbf{U} + \operatorname{grad} s = \lambda \mathbf{w} \text{ in } \Omega, \tag{15.3}$$

$$\text{div } \mathbf{w} = 0 \text{ in } \Omega, \tag{15.4}$$

and

$$\mathbf{w} = \mathbf{0} \text{ on } \Gamma. \tag{15.5}$$

Then the given flow field $\mathbf{U}$ is said to be *asymptotically stable to infinitesimal perturbations* if $\mathfrak{Re}(\lambda) > 0$ *for all eigenvalues* λ of (15.3)–(15.5). See Drazin and Reid [1981] for detailed discussions of the relation of (15.3)–(15.5) to linearized stability analyses.

Once again, except for some simple cases, one cannot in general determine the eigenvalues of (15.3)–(15.5), except through numerical procedures. As usual, in order to define finite element approximations for the eigenvalues of (15.3)–(15.5), we recast the latter into a weak form. Unfortunately, due to the fact that (15.3)–(15.5) is not self-adjoint, one must expect (15.3)–(15.5) to possess complex valued eigenvalues and eigenvectors. Thus, given $\mathbf{U} \in \mathbf{H}'(\Omega)$, we now seek $0 \neq \mathbf{w} \in \mathbf{H}_0^1(\Omega)$, $0 \neq s \in L_0^2(\Omega)$, and $\lambda \in \mathbb{C}$ such that

$$a(\mathbf{w}, \mathbf{v}) + c(\mathbf{U}, \mathbf{w}, \mathbf{v}) + c(\mathbf{w}, \mathbf{U}, \mathbf{v}) + b(\mathbf{v}, s)$$

$$= \lambda(\mathbf{w}, \mathbf{v}) \quad \text{for all } \mathbf{v} \in \mathbf{H}_0^1(\Omega) \tag{15.6}$$

$$\overline{b(\mathbf{w}, q)} = 0 \quad \text{for all } q \in L_0^2(\Omega), \tag{15.7}$$

where $(\overline{\cdot})$ denotes complex conjugation. Here $\mathbf{H}'(\Omega)$, $\mathbf{H}_0^1(\Omega)$, and $L_0^2(\Omega)$ denote spaces of *complex valued* functions and, for given real $\mathbf{U}$, $a(\cdot, \cdot)$, $b(\cdot, \cdot)$, $c(\cdot, \mathbf{U}, \cdot)$, and $c(\mathbf{U}, \cdot, \cdot)$ are *sesquilinear* forms whose definitions are the obvious generalizations of (1.6)–(1.8) to the complex case. For example, now we have that

$$b(\mathbf{v}, q) = \int_\Omega q \, \text{div } \bar{\mathbf{v}} \, d\Omega \quad \text{and} \quad c(\mathbf{w}, \mathbf{U}, \mathbf{v}) = \int_\Omega \mathbf{w} \cdot \text{grad } \mathbf{U} \cdot \bar{\mathbf{v}} \, d\Omega.$$

15.2. Finite Element Approximations

Finite element approximations are defined in the usual manner. One chooses finite dimensional (complex) subspaces $\mathbf{V}_0^h \subset \mathbf{H}_0^1(\Omega)$ and $S_0^h \subset L_0^2(\Omega)$ and then, given real $\mathbf{U} \in \mathbf{H}'(\Omega)$, one seeks $\mathbf{w}^h \in \mathbf{V}_0^h$, $s^h \in S_0^h$, and $\lambda^h \in \mathbb{C}$ such that

$$a(\mathbf{w}^h, \mathbf{v}^h) + c(\mathbf{U}, \mathbf{w}^h, \mathbf{v}^h) + c(\mathbf{w}^h, \mathbf{U}, \mathbf{v}^h) + b(\mathbf{v}^h, s^h)$$

$$= \lambda^h(\mathbf{w}^h, \mathbf{v}^h) \quad \text{for all } \mathbf{v}^h \in \mathbf{V}_0^h \tag{15.8}$$

and

$$\overline{b(\mathbf{w}^h, q^h)} = 0 \qquad \text{for all } q^h \in S_0^h. \tag{15.9}$$

Once again we assume that the approximating subspaces satisfy all the conditions, extended to the complex case, necessary for the stable approximation of the solutions of the Navier–Stokes equations. Then, one may prove the following results, which were obtained in Osborn [1976]. Let λ denote any eigenvalue of (15.6)–(15.7) and let m denote its algebraic multiplicity. Then there are exactly m eigenvalues of (15.8)–(15.9), counted according to algebraic multiplicites, that converge to λ. Denote these m eigenvalues of (15.8)–(15.9) by λ_j^h, $j = 1, \ldots, m$ and let

$$\tilde{\lambda}^h = \left(\sum_{j=1}^{m} \frac{1}{\lambda_j^h} \right)^{-1}.$$

Then

$$|\lambda - \tilde{\lambda}^h| = O(h^{2\ell}) \tag{15.10}$$

whenever $\mathbf{w} \in \mathbf{H}^{\ell+1}(\Omega) \cap \mathbf{H}_0^1(\Omega)$ and $s \in H^\ell \cap L_0^2(\Omega)$ and where, e.g., $\ell = 1$ for the finite element spaces of Section 3.1 and $\ell = 2$ for the Taylor–Hood element of Secton 3.2. Due to the fact that (15.3)–(15.5), (15.6)–(15.7), and (15.8)–(15.9) are not self-adjoint, one cannot guarantee, unless the geometric and algebraic multiplicities of λ are the same, that the individual eigenvalues λ_j^h, $j = 1, \ldots, m$, converge at the rates indicated by (15.10). Instead, we have that

$$|\lambda - \lambda_j^h| = O(h^{2\ell/\alpha}),$$

where ℓ is as in (15.10) and α is the *ascent* of the eigenvalue λ. For details concerning the ascent, see, e.g., Osborn [1976]. Here it suffices to say that in general $1 \le \alpha \le m$, where m is the algebraic multiplicity of λ. Indeed, $\alpha = 1$ whenever the algebraic and geometric multiplicities of λ are equal, i.e., whenever there are m linearly independent eigenvectors corresponding to λ. On the other hand, if the geometric multiplicity is unity, i.e., there is only one linearly independent eigenvector corresponding to the eigenvalue λ, then $\alpha = m$. When the geometric multiplicity is strictly between 1 and m, the ascent is determined by examining the generalized eigenspaces corresponding to λ. For example, in the finite dimensional case, the ascent is simply the dimension of the largest Jordan block associated with the eigenvalue λ.

VII

EXTERIOR PROBLEMS

Many flow simulations are modeled through mathematical problems posed in exterior domains. Rigorous studies of such problems have been largely confined to linear problems, and thus we only consider the Stokes problem. However, although the theoretical results we will discuss have been rigorously demonstrated only in the linear case, from an *algorithmic* point of view, the methods discussed here can be applied to the nonlinear Navier–Stokes case as well. We do not discuss methods whose definitions depend in a crucial manner, e.g., by using free space Green's functions, on the linearity of the Stokes equation.

There are a variety of approaches that one may take in attacking problems posed on unbounded domains. One method is to map, perhaps conformally, the domain into one of bounded extent. This method often works well and is attractive from an analytical point of view. Of course, the price paid is that the linear *constant coefficient* Stokes equations are transformed, in the mapped domain, into a linear *variable coefficient* system of partial differential equations, albeit, posed on a finite domain. Furthermore, for general boundaries, the mapping of the exterior domain into a bounded one must be done numerically, often at considerable expense.

A finite number of degrees of freedom may also be obtained through the use of a discretization that uses finite elements of different sizes, including ones of infinite extent. This approach has been analyzed in contexts other than the Stokes equations by Goldstein [1981]. It requires substantially different coding, e.g., quadrature rules and basis functions, than do problems posed on bounded domains.

Another approach is to replace the exterior domain problem with a problem posed on a bounded domain, usually requiring the same partial differential equations to be satisfied in the truncated domain. This necessitates the imposition of a boundary condition on the artificial boundary introduced by the truncation process. At least for the linear Stokes problem, boundary conditions may be devised such that, within the truncated domain, the solution of the truncated domain problem is the same as that of the original exterior problem. Unfortunately, such an artificial boundary condition is necessarily *nonlocal*, i.e., instead of holding pointwise along the artificial boundary, it couples together all points on the boundary. Among other things, this coupling complicates the discretization and solution processes connected with, e.g., finite element methods.

Local artificial boundary conditions may be used if one is willing to accept solutions of the truncated domain problem that are not exactly the same, even within the truncated domain, as that of the original exterior domain problem. Since our aim is to find approximate solutions via finite element methods, we will not be finding exact solutions anyway, and therefore this approach is, from a computational point of view, the most attractive. For example, a bounded domain finite element code for the Navier–Stokes equation may be easily converted into one that treats exterior problems through a truncated domain/ artificial boundary condition approach.

Here we exclusively consider the last approach, following closely Guirguis [1986 and 1987a] and Guirguis and Gunzburger [1987]. Furthermore, we will only look at problems posed in exterior domains of $\mathbb{R}^3$. Coupled boundary integral/finite element methods are considered by Guirguis [1987b] and Sequeira [1983 and 1986]. See Guirguis [1988] for a demonstration of the superiority of the truncated domain/ artificial boundary condition approach compared to the coupled boundary integral/finite element method. Other possible approaches, discussed in the context of the Laplace, Helmholtz, and other equations, may be found in Bayliss, Gunzburger, and Turkel [1982], Cantor [1983], LeRoux [1977], Nedelec and Planchard [1973], and O'Leary and Widlund [1979].

16

Truncated Domain-
Artificial Boundary
Condition Methods

16.1. Truncated Domain Problems

Let Ω_1 denote a bounded domain of $\mathbb{R}^3$ with boundary Γ and let Ω denote the complement, in $\mathbb{R}^3$, of $\bar{\Omega}_1$. The domain Ω_1 may be empty, in which case $\Omega = \mathbb{R}^3$. In order for the analyses discussed below to be valid, we need to make the technical assumption that whenever Ω_1 is not empty, it is star-shaped with respect to the origin, i.e., points on any line joining the origin to the boundary Γ belong to $\bar{\Omega}_1$. The goal is to find approximations of the solution of the exterior Stokes problem

$$-\mathrm{div}(\mathrm{grad}\,\mathbf{u} + (\mathrm{grad}\,\mathbf{u})^T) + \mathrm{grad}\,p = \mathbf{f} \text{ in } \Omega, \qquad (16.1)$$

$$\mathrm{div}\,\mathbf{u} = 0 \text{ in } \Omega, \qquad (16.2)$$

$$\mathbf{u} = \mathbf{0} \text{ on } \Gamma \qquad (16.3)$$

and

$$\lim_{|\mathbf{x}|\to\infty} \mathbf{u}(\mathbf{x}) \to \mathbf{0}. \qquad (16.4)$$

As usual, $\mathbf{u}$, p, and $\mathbf{f}$ respectively denote the velocity, pressure, and body force; the constant density and viscosity coefficient have been absorbed into p and $\mathbf{f}$; and $\mathbf{x}$ denotes a point of $\mathbb{R}^3$. Boundary conditions other than (16.3) may also be treated as long as the analysis of approximations of an analogous Stokes problem posed on a bounded domain has been carried out.

In order to avoid the infinite exterior domain of (16.1)–(16.4), we instead consider the *truncated domain problem*

$$-\text{div}(\text{grad}\,\mathbf{u}_R + (\text{grad}\,\mathbf{u}_R)^T) + \text{grad}\,p_R = \mathbf{f}_R = \mathbf{f}\big|_{\Omega_R} \text{ in } \Omega_R, \qquad (16.5)$$

$$\text{div}\,\mathbf{u}_R = 0 \text{ in } \Omega_R, \qquad (16.6)$$

$$\mathbf{u}_R = \mathbf{0} \text{ on } \Gamma, \qquad (16.7)$$

and

$$\mathcal{B}_i(\mathbf{u}_R, p_R) = 0 \text{ on } \Gamma_R, \quad i = 1 \text{ or } 2. \quad (16.8)$$

In (16.5)–(16.8), Ω_R and Γ_R denote the truncated domain and the artificial boundary. For example, if $B(0, R)$ denotes the ball of radius R centered at the origin and $\partial B(0, R)$ denotes its surface, i.e., the sphere of radius R centered at the origin, we could have that

$$\Omega_R = \Omega \cap B(0, R) \quad \text{and} \quad \Gamma_R = \partial B(0, R).$$

In general, we will assume that $\Omega_R \subset B(0, R)$. Equation (16.8) denotes the artificial boundary condition that is imposed in order to close the truncated domain problem. We will consider two choices for this artificial boundary condition, namely,

$$\mathcal{B}_1(\mathbf{u}_R, p_R) = \mathbf{u}_R = \mathbf{0} \qquad (16.9)$$

and

$$\mathcal{B}_2(\mathbf{u}_R, p_R) = -p_R\,\mathbf{n} + \mathbf{n} \cdot (\text{grad}\,\mathbf{u}_R + (\text{grad}\,\mathbf{u}_R)^T) + \frac{\delta}{R}\mathbf{u}_R = \mathbf{0}, \quad (16.10)$$

where $\delta = 1$ if Ω_1 is empty, and $\delta = 0$ otherwise. A third, and higher order, choice for the artificial boundary condition is analyzed in Guirguis [1987a].

16.2. Function Spaces

In order to enable the analysis of the exterior problem (16.1)–(16.4) and the truncated problem (16.5)–(16.8), we introduce the weighted Sobolev spaces of Hanouzet [1971]. Furthermore, by using these spaces we are able to keep precise track of how constants appearing in the various error estimates depend on R.

Let $\lambda = (\lambda_1, \lambda_2, \lambda_3)$ for λ_i, $i = 1, 2, 3$, non-negative integers be a multi-index and let $|\lambda| = \lambda_1 + \lambda_2 + \lambda_3$. For m a non-negative integer, α any real number, and $\mathfrak{D}$ a possibly unbounded set in $\mathbb{R}^3$, we define

$$U_\alpha^m(\mathfrak{D}) = \left\{ q : \int_{\mathfrak{D}} (1 + |\mathbf{x}|^2)^{\alpha - m + |\lambda|} |D^\lambda q|^2 \, d\Omega < \infty, \quad |\lambda| \leq m \right\}.$$

The inclusion of the one in the weight $(1 + |\mathbf{x}|^2)$ is necessary in case $\mathfrak{D} = \mathbb{R}^3$, in which case $\mathfrak{D}$ contains the origin. We briefly mention some of the properties of these spaces; details may be found in Hanouzet [1971] and Guirguis [1986]. First, these are Hilbert spaces equipped with the inner product

$$(q, p)_{m, \alpha, \mathfrak{D}} = \sum_{|\lambda| \leq m} \int_{\mathfrak{D}} (1 + |\mathbf{x}|^2)^{\alpha - m + |\lambda|} |D^\lambda q| \, |D^\lambda p| \, d\Omega$$

and the norm

$$\|q\|_{m, \alpha, \mathfrak{D}} = [(q, q)_{m, \alpha, \mathfrak{D}}]^{1/2}.$$

Of particular interest will be the spaces $U_0^0(\mathfrak{D}) = L^2(\mathfrak{D})$ and

$$U_0^1(\mathfrak{D}) = \left\{ q : \int_{\mathfrak{D}} \left\{ \sum_{|\lambda| = 1} |D^\lambda q|^2 + \frac{q^2}{(1 + |\mathbf{x}|^2)} \right\} d\Omega < \infty \right\}.$$

We will also need to make use of the spaces

$$\mathring{L}^2(\mathfrak{D}) = \left\{ q \in L^2(\mathfrak{D}) : \int_{\mathfrak{D}} q \, d\Omega = 0 \right\}$$

and

$$\mathring{U}_0^1(\mathfrak{D}) = \{ q \in U_0^1(\mathfrak{D}) : q = 0 \text{ on } \partial\mathfrak{D} \},$$

and the dual space $U_0^{-1}(\mathfrak{D})$, which is equipped with the norm

$$\|q\|_{-1, 0, \mathfrak{D}} = \sup_{0 \neq p \, \in \mathring{U}_0^1(\mathfrak{D})} \frac{\int_{\mathfrak{D}} qp \, d\Omega}{\|p\|_{1, 0, \mathfrak{D}}}.$$

Spaces, inner products, and norms for vector valued functions may be defined from the above spaces in the usual manner. We will denote, for example, by $\mathbf{U}_0^1(\mathfrak{D})$ the space of vector valued functions each of whose components belong to $U_0^1(\mathfrak{D})$.

16.3. Weak Formulations and Error Estimates for the Truncated Problems

We introduce the bilinear forms

$$\hat{a}(\mathbf{u}, \mathbf{v}) = \frac{1}{2} \int_{\Omega} (\operatorname{grad} \mathbf{u} + (\operatorname{grad} \mathbf{u})^T) : (\operatorname{grad} \mathbf{v} + (\operatorname{grad} \mathbf{v})^T) \, d\Omega$$

$$\text{for } \mathbf{u}, \mathbf{v} \in \mathbf{U}_0^1(\Omega), \tag{16.11}$$

and

$$b(\mathbf{v}, q) = -\int_{\Omega} q \operatorname{div} \mathbf{v} \, d\Omega \quad \text{for } \mathbf{v} \in \mathbf{U}_0^1(\Omega) \text{ and } q \in L^2(\Omega), \tag{16.12}$$

and the linear functional

$$\mathcal{F}(\mathbf{v}) = \int_{\Omega} \mathbf{f} \cdot \mathbf{v} \, d\Omega \quad \text{for } \mathbf{f} \in \mathbf{U}_0^{-1}(\Omega) \text{ and } \mathbf{v} \in \mathbf{U}_0^1(\Omega). \tag{16.13}$$

Then a variational form of (16.1)–(16.4) is given as follows: Given $\mathbf{f} \in \mathbf{U}_0^{-1}(\Omega)$, seek $\mathbf{u} \in \mathring{\mathbf{U}}_0^1(\Omega)$ and $p \in L^2(\Omega)$ such that

$$\hat{a}(\mathbf{u}, \mathbf{v}) + b(\mathbf{v}, p) = \mathcal{F}(\mathbf{v}) \quad \text{for all } \mathbf{v} \in \mathring{\mathbf{U}}_0^1(\Omega) \tag{16.14}$$

$$b(\mathbf{u}, q) = 0 \quad \text{for all } q \in L^2(\Omega). \tag{16.15}$$

The existence and uniqueness of solutions of (16.14)–(16.15), as well as regularity results, are established in Guirguis [1986]. Note that this weak formulation does not use any weights in the integrals; however, due to the weights appearing in the definition of the relevant spaces, the forms (16.11)–(16.13) are all well defined on these function spaces. Note that since constant functions do not belong to $L^2(\Omega)$ when Ω is an exterior domain, we do not need to constrain the pressure space.

The first truncated problem will make use of the bilinear forms

$$a_{1R}(\mathbf{u}, \mathbf{v}) = \frac{1}{2} \int_{\Omega_R} (\operatorname{grad} \mathbf{u} + (\operatorname{grad} \mathbf{u})^T) : (\operatorname{grad} \mathbf{v} + (\operatorname{grad} \mathbf{v})^T) \, d\Omega$$

$$\text{for } \mathbf{u}, \mathbf{v} \in \mathbf{U}_0^1(\Omega_R), \tag{16.16}$$

and

$$b_R(\mathbf{v}, q) = -\int_{\Omega_R} q \operatorname{div} \mathbf{v} \, d\Omega \quad \text{for } \mathbf{v} \in \mathbf{U}_0^1(\Omega_R) \text{ and } q \in L^2(\Omega_R) \tag{16.17}$$

and the linear functional

$$\mathfrak{F}_R(\mathbf{v}) = \int_{\Omega_R} \mathbf{f} \cdot \mathbf{v} \, d\Omega \quad \text{for } \mathbf{f} \in \mathbf{U}_0^{-1}(\Omega) \text{ and } \mathbf{v} \in \mathbf{U}_0^1(\Omega_R). \quad (16.18)$$

Given $\mathbf{f} \in \mathbf{U}_0^{-1}(\Omega)$, we now seek $\mathbf{u}_R \in \mathbf{V}_1 = \mathring{\mathbf{U}}_0^1(\Omega_R)$ and $p_R \in S_1 = \mathring{L}^2(\Omega_R)$ such that

$$a_{1R}(\mathbf{u}_R, \mathbf{v}) + b_R(\mathbf{v}, p_R) = \mathfrak{F}_R(\mathbf{v}) \quad \text{for all } \mathbf{v} \in \mathbf{V}_1 \quad (16.19)$$

and

$$b_R(\mathbf{u}_R, q) = 0 \quad \text{for all } q \in S_1. \quad (16.20)$$

The weak formulation (16.19)–(16.20) contains the first artificial boundary condition (16.9) as an essential boundary condition. Also, we are implicitly using the fact that $\mathbf{f}_R = \mathbf{f}|_{\Omega_R}$. Note that since Ω_R is a bounded domain, we now need to constrain the pressure space in order to remove constant functions.

For the second truncated problem, we use the space

$$\mathbf{V}_2 = \{\Phi_R \mathbf{v} : \mathbf{v} \in \mathring{\mathbf{U}}_0^1(\Omega)\} = \{\mathbf{v} \in \mathbf{H}^1(\Omega_R) : \mathbf{v} = \mathbf{0} \text{ on } \Gamma\},$$

where Φ_R denotes the characteristic function of the domain Ω_R and the bilinear form

$$a_{2R}(\mathbf{u}, \mathbf{v}) = a_{1R}(\mathbf{u}, \mathbf{v}) + \frac{\delta}{R} \int_{\Gamma_R} \mathbf{u} \cdot \mathbf{v} \, d\Gamma \quad \text{for } \mathbf{u}, \mathbf{v} \in \mathbf{V}_2. \quad (16.21)$$

The last term in (16.21) is included in order to render the form $a_{2R}(\cdot, \cdot)$ coercive on $\mathbf{V}_2 \times \mathbf{V}_2$. In case Ω_1 is not empty, that term may be omitted. Then, given $\mathbf{f} \in \mathbf{U}_0^{-1}(\Omega)$, we now seek $\mathbf{u}_R \in \mathbf{V}_2$ and $p_R \in S_2 = L^2(\Omega_R)$ such that

$$a_{2R}(\mathbf{u}_R, \mathbf{v}) + b_R(\mathbf{v}, p_R) = \mathfrak{F}_R(\mathbf{v}) \quad \text{for all } \mathbf{v} \in \mathbf{V}_2 \quad (16.22)$$

and

$$b_R(\mathbf{u}_R, q) = 0 \quad \text{for all } q \in S_2, \quad (16.23)$$

where $b_R(\cdot, \cdot)$ and $\mathfrak{F}_R(\cdot)$ are given by (16.17) and (16.18), respectively. The second artificial boundary condition (16.10) is natural to this weak formulation. Again, we are implicitly using the fact that $\mathbf{f}_R = \mathbf{f}|_{\Omega_R}$. In (16.22)–(16.23) the pressure space is not constrained due to the appearance of the pressure in the natural boundary condition (16.10).

The existence and uniqueness of the solutions of (16.19)–(16.20) and (16.22)–(16.23), as well as regularity results, are established in

Guirguis [1986]. Moreover, the following estimate for the difference between the solution of the truncated problems and that of the exterior problem (16.14)–(16.15) is also derived in that reference. Suppose that as $R \to \infty$ the data $\mathbf{f} = O(R^{-r})$ for some $r > 1/2$. Then

$$\|\mathbf{u}_R - \mathbf{u}\|_{1,0,\Omega_R} + \|p_R - p\|_{0,0,\Omega_R} \le C(R^{-r+1/2} + R^{-i+1/2}), \quad (16.24)$$

where C is a constant independent of R and $i = 1$ or 2, corresponding to the two possible choices of artificial boundary conditions (16.9) or (16.10), or equivalently, to the two weak formulations (16.19)–(16.20) or (16.22)–(16.23). If $\mathbf{f}$ has compact support in Ω_R, then the first term on (16.24) vanishes, i.e.,

$$\|\mathbf{u}_R - \mathbf{u}\|_{1,0,\Omega_R} + \|p_R - p\|_{0,0,\Omega_R} \le CR^{-i+1/2}. \quad (16.25)$$

Thus, for example, (16.25) indicates errors of $O(R^{-1/2})$ and $O(R^{-3/2})$ for the truncated problems (16.19)–(16.20) and (16.22)–(16.23), respectively. In Guirguis [1987a] a more accurate boundary condition is studied, for which the estimates (16.24) and (16.25) hold with $i = 3$.

It is important to note that the norms used in (16.24) and (16.25) to measure the error in the solution of the truncated problem have leading terms that do not involve any weights, i.e., $\|\cdot\|_{0,0,\Omega_R}$ is simply the ordinary $L^2(\Omega_R)$-norm $\|\cdot\|_{0,\Omega_R}$ and the term involving first derivatives in $\|\cdot\|_{1,0,\Omega_R}$ is exactly the ordinary $H^1(\Omega_R)$-semi-norm $|\cdot|_{1,\Omega_R}$. Therefore, for example, (16.24) implies that

$$|\mathbf{u}_R - \mathbf{u}|_{1,\Omega_R} + \|p_R - p\|_{0,\Omega_R} \le C(R^{-r+1/2} + R^{-i+1/2}), \quad (16.26)$$

where we are now employing only standard unweighted Sobolev norms to measure errors in the solution of the truncated problems.

16.4.   Finite Element Approximations

We now consider the approximation of the solution of the truncated problem by finite element methods. Since Ω_R is by construction a bounded domain, the definition of finite element algorithms is straightforward. However, in analyzing the error in these algorithms, it is imperative to use the weighted spaces of Hanouzet introduced above in order to derive error estimates containing constants that are independent of the truncation parameter R.

To define finite element methods we proceed in the usual manner. We choose finite dimensional subspaces $\mathbf{V}^h \subset \mathbf{V}_i$ and $S^h \subset S_i$, $i = 1$ or 2, depending on which artificial boundary condition is being used. We then seek $\mathbf{u}^h \in \mathbf{V}^h$ and $p^h \in S^h$ such that, for $i = 1$ or 2,

$$a_{iR}(\mathbf{u}^h, \mathbf{v}) + b_R(\mathbf{v}, p^h) = \mathcal{F}_R(\mathbf{v}) \quad \text{for all } \mathbf{v} \in \mathbf{V}^h \tag{16.27}$$

and

$$b_R(\mathbf{u}^h, q) = 0 \quad \text{for all } q \in S^h. \tag{16.28}$$

Since Ω_R is a bounded domain, the weighted spaces $U_0^1(\Omega_R)$ and $U_0^0(\Omega_R)$ are equivalent to the standard unweighted spaces $H^1(\Omega_R)$ and $L^2(\Omega_R)$, respectively. Thus, all results, including those concerning div-stability, discussed in Chapters 1–5 hold here as well. In particular, any pair of velocity and pressure finite element spaces that was found to be stable for the problems discussed in Chapters 1–5 may be used here to obtain stable discretizations of the truncated domain problems.

We make the following standard assumption concerning the ability to approximate in the finite element spaces (Ciarlet [1978]). Suppose that $\mathbf{v} \in U_m^{m+1}(\Omega_R) \approx H^{m+1}(\Omega_R)$ and $q \in U_m^m(\Omega_R) \approx H^m(\Omega_R)$, then there exist $\mathbf{v}^h \in \mathbf{V}^h$, $q^h \in S^h$, and a constant C independent of h and R such that

$$\|\mathbf{v}^h - \mathbf{v}\|_{1,\Omega_R} + \|q^h - q\|_{0,\Omega_R} \leq Ch^m(|\mathbf{v}|_{m+1,\Omega_R} + |q|_{m,\Omega_R}), \tag{16.29}$$

where the norms and semi-norms are all of the unweighted type. For example, for the Taylor–Hood element pair (16.29), holds with $m = 2$.

Now assume that the finite element spaces $\mathbf{V}^h$ and S^h satisfy the div-stability property of Chapter 2, with respect to the domain Ω_R, and also satisfy the approximation property (16.29). Also assume that $\mathbf{u} \in U_m^{m+1}(\Omega_R) \approx H^{m+1}(\Omega_R)$ and $p \in U_m^m(\Omega_R) \approx H^m(\Omega_R)$. Then

$$\|\mathbf{u}^h - \mathbf{u}_R\|_{1,0,\Omega_R} + \|p^h - p_R\|_{0,0,\Omega_R} \leq Ch^m,$$

with a constant C independent of both h and R. Using the same reasoning that took us from (16.24) to (16.26) then implies that

$$|\mathbf{u}^h - \mathbf{u}_R|_{1,\Omega_R} + \|p^h - p_R\|_{0,\Omega_R} \leq Ch^m, \tag{16.30}$$

where C is still independent of h and R.

Combining (16.30) with (16.26) yields an error estimate involving the truncation parameter R and the discretization parameter h. We summarize the above results by giving a precise statement of this estimate. Let $\mathbf{u}, p$ and $\mathbf{u}^h, p^h$ denote the solutions of the exterior problem (16.14)–(16.15) and of the discrete finite element problem (16.27)–(16.28). Suppose $\mathbf{u} \in U_m^{m+1}(\Omega)$ and $p \in U_m^m(\Omega)$ for some positive integer m and that the finite element spaces $\mathbf{V}^h$ and S^h satisfy the div-stability condition of Chapter 2 and the approximating property (16.29). Also suppose that as $R \to \infty$ the data $\mathbf{f} = O(R^{-r})$ for some $r > 1/2$. Then there exists a constant C, independent of both h and R, such that

$$|\mathbf{u}^h - \mathbf{u}|_{1,\Omega_R} + \|p^h - p\|_{0,\Omega_R} \le C(R^{-r+1/2} + R^{-i+1/2} + h^m), \quad (16.31)$$

where $i = 1$ or 2 according to which of the two artificial boundary conditions (16.9) or (16.10) is used. If, on the other hand, $\mathbf{f}$ has compact support, then for R sufficiently large

$$|\mathbf{u}^h - \mathbf{u}|_{1,\Omega_R} + \|p^h - p\|_{0,\Omega_R} \le C(R^{-i+1/2} + h^m). \quad (16.32)$$

We again emphasize that although the final estimates (16.31) and (16.32) involve unweighted norms, their derivation requires the use of weighted norms in order to keep track of dependences on the truncation parameter R.

For the sake of efficiency in practical implementations, it is useful to choose h and R so that the discretization and truncation error terms appearing in the estimates (16.31) or (16.32) are balanced. However, it should be pointed out that a quasi-uniform triangulation of Ω_R will not take advantage of the fact that the solution of the exterior problem decays as $|\mathbf{x}| \to \infty$. This decay property allows, without adversely affecting the accuracy, for the use of graded meshes in which the size of the elements grows as one triangulates further away from the origin. The use of such meshes results, of course, in substantially less degrees of freedom, compared to using quasi-uniform meshes. Details concerning the use of graded meshes may be found in Goldstein [1981].

There is computational evidence that the estimates (16.31) and (16.32) are sharp. Moreover, there is also computational evidence that suggests that if we measure errors on *fixed* subdomains we cannot improve on the exponent of R, i.e., if we let $\mathfrak{D} \subset \Omega_R$ for all R we will have that

$$|\mathbf{u}^h - \mathbf{u}|_{1,\mathfrak{D}} + \|p^h - p\|_{0,\mathfrak{D}} \approx C(R^{-r+1/2} + R^{-i+1/2} + h^m),$$

where C is still independent of R and h.

VIII

NONLINEAR CONSTITUTIVE RELATIONS

Our aim in this part is to consider the approximation, by finite element methods, of solutions of some alternate models for incompressible flow that feature nonlinear constitutive laws. There have been many such models proposed; we have chosen two that have undergone considerable analyses within the finite element framework. An important class of models that we do not consider here are the Maxwell (and related) models for viscoelastic flow. For these models, the constitutive relation takes the form of differential equations relating the components of the stress tensor to those of the velocity field. One may consult Crochet and Walters [1983], Crochet, Davies, and Walters [1984], Marchal and Crochet [1987], and van Schaftingen and Crochet [1984] for discussions of finite element algorithms for these types of models.

17

A Ladyzhenskaya
Model and Algebraic
Turbulence Models

17.1. A Ladyzhenskaya Model and Its Relation
to Algebraic Turbulence Models

The Navier–Stokes equations (1.1) are derived from the principle of conservation of linear momentum and the use of the linear constitutive relation

$$\sigma = -pI + \mu(\text{grad } \mathbf{u} + (\text{grad } \mathbf{u})^T), \tag{17.1}$$

where I, σ, p, $\mathbf{u}$, and μ denote the identity tensor, the stress tensor, the pressure, the velocity, and the dynamic viscosity coefficient, respectively. For a variety of physical, mathematical, and computational reasons, this constitutive law is sometimes replaced by a *nonlinear* constitutive relation.

It has not been shown, nor is it believed, that the Navier–Stokes equations have globally (in time) unique solutions in three-dimensional settings. This is perhaps in conflict with physical intuition, through which one may expect that physical flows, once the boundary and initial data are specified, are uniquely determined. In order to resolve this discrepancy, Ladyzhenskaya [1969, 1970a, and 1970b] proposed some alternate models for viscous incompressible flow, all of which possess globally unique solutions.

The Navier–Stokes model may be viewed as a special case of the Stokesian model for a fluid (Serrin [1959]) in which only linear terms in the stress/rate of the strain constitutive relation are retained. The particular Ladyzhenskaya model considered here, at least for some choices of parameters, may also be viewed as a special case of the Stokesian model, where now *some* nonlinear terms are kept. Since we are interested in finite element discretizations, we will only consider a stationary model. For analyses concerned with both the stationary and time-dependent equations and their solutions, see Du and Gunzburger [1989a] and Ladyzhenskaya [1970a and 1970b].

The particular model we consider is given by

$$-\text{div}(\mathcal{Q}(\mathbf{u})[\text{grad } \mathbf{u} + (\text{grad } \mathbf{u})^T]) + \mathbf{u} \cdot \text{grad } \mathbf{u} + \text{grad } p = \mathbf{f} \text{ in } \Omega, \quad (17.2)$$

$$\text{div } \mathbf{u} = 0 \text{ in } \Omega, \quad (17.3)$$

and

$$\mathbf{u} = \mathbf{0} \text{ on } \Gamma, \quad (17.4)$$

where as usual $\mathbf{u}$, p, and $\mathbf{f}$ denote the velocity, pressure, and body force, respectively, and Ω denotes an open, bounded subset of $\mathbb{R}^n$, $n = 2$, or 3, with boundary Γ. In (17.2) we have that

$$\mathcal{Q}(\mathbf{u}) = \nu + \nu_1 \left[\sum_{i,j=1}^{n} \left(\frac{\partial u_i}{\partial x_j} \right)^2 \right]^{r/2}, \quad (17.5)$$

where ν may be interpreted as the usual kinematic viscosity, $r > 0$ is a constant, and ν_1 is a given function. Some information concerning the values of r and the functional form of ν_1 may be gleaned from kinetic theory (Ladyzhenskaya [1970a and 1970b]); in general $\nu_1 > 0$ and $\nu_1 \ll \nu$.

Clearly, if we take $\nu_1 = 0$, then (17.2) reduces to the Navier–Stokes equation. Also, for small shear rates, i.e., when the derivatives of the velocity are small, again $\mathcal{Q}(\mathbf{u}) \approx \nu$ and (17.2) essentially reduces to the Navier–Stokes equation. However, for large shear rates, the additional terms in the "viscosity" $\mathcal{Q}(\mathbf{u})$ result in a more highly damped system. Thus it is not surprising that Ladyzhenskaya was able to prove that solutions of the time-dependent version of (17.2)–(17.4) possess globally unique solutions regardless of the size of the data $\mathbf{f}$.

The constitutive relation inherent to (17.2) is

$$\sigma = -pI + \left\{ \rho\nu + \rho\nu_1 \left[\sum_{i,j=1}^{n} \left(\frac{\partial u_i}{\partial x_j} \right)^2 \right]^{r/2} \right\} (\text{grad } \mathbf{u} + (\text{grad } \mathbf{u})^T), \quad (17.6)$$

where ρ denotes the constant density. As noted above, this relation may be viewed for certain values of r, e.g., $r = 2$, as describing a more general Stokesian fluid than that described by the linear constitutive relation (17.1). However, there is another modeling avenue through which one may arrive at (17.6) as a constitutive relation for a fluid, namely, when one uses algebraic turbulence models. Here one retains the Navier–Stokes equations in the basic mathematical model for turbulent fluid motion but recognizes that it is impossible, using today's computers, to resolve the "small"-scale phenomena present in such motions. Thus one only solves for "larger"-scale structures, i.e., those resolvable using grid sizes that can, in practice, be handled with present-day computing power.

The equations governing these "larger"-scale structures are found by taking temporal and/or spatial averages of the Navier–Stokes equations. This process results in equations involving not only the average velocity and pressure, but also additional degrees of freedom, e.g., the Reynolds stresses, arising from the average of quadratic combinations of the velocity components and their derivatives. These additional terms account for the effects of the neglected "small"-scale phenomena on the "large"-scale structures. However, in order to have a closed, solvable system one must eliminate these additional degrees of freedom, i.e., one must somehow relate them to the average velocity. This latter procedure is known as "turbulence modeling" and gives rise to a variety of models, including algebraic or zero-equation models, k or one-equation models, $k - \varepsilon$ or two-equation models, large eddy models, group renormalization models, etc.; see any book on turbulence, e.g., Bradshaw, Cebeci, and Whitelaw [1981]; Hinze [1959]; Stanisic [1985]; and Tennekes and Lumley [1987], for a derivation and description of these models.

Algebraic turbulence models are those for which the additional degrees of freedom arising from the averaging procedure are related to the derivatives of the velocity, i.e., to the rate of strain, through an *algebraic* relation. A popular choice for this last relation results in (17.2)–(17.5) with $r = 2$ where $\mathbf{u}$ and p now denote the average velocity and pressure, respectively. When (17.2)–(17.5) are viewed as an algebraic turbulence model, the function v_1 is proportional to an empirical function known as the *mixing length*.

Algebraic turbulence models are also known as *zero equation* models because no additional differential equations need be solved in order to relate the Reynolds stresses to the average velocity. This is in contrast to, e.g., the k and $k - \varepsilon$ models, which involve one and two, respectively, additional differential equations. A variety of finite element algorithms for k and $k - \varepsilon$ models have been proposed in the engineering literature; however, no rigorous analyses of these have been carried out, and we will not consider them further. Algebraic turbulence models and their finite element discretization are considered in Gunzburger and Turner [1988] and Turner [1989], and are included in the analyses of Du and Gunzburger [1989a and 1989b].

17.2. Finite Element Approximations

We define the space

$$\mathbf{V}_0 = \{\mathbf{v} \in \mathbf{W}^{1,r+2}(\Omega) : \mathbf{v} = \mathbf{0} \text{ on } \Gamma\},$$

where as usual $\mathbf{W}^{1,s}(\Omega)$ denotes the Sobolev space consisting of vector valued functions with components having all first derivatives belonging to $L^s(\Omega)$. We denote by $\mathbf{V}'$ the dual space of $\mathbf{V}_0$ with the duality pairing between $\mathbf{V}'$ and $\mathbf{V}_0$ being defined through the $L^2(\Omega)$ inner product. The seminorm

$$|\mathbf{v}|_{1,s} = \left(\int_\Omega |\text{grad } \mathbf{v}|^s \, d\Omega \right)^{1/s},$$

$s = r + 2$, defines a norm on $\mathbf{V}_0$.

A weak formulation of (17.2)–(17.5) is then given as follows. Given $\mathbf{f} \in \mathbf{V}'$, find $\mathbf{u} \in \mathbf{V}_0$ and $p \in S_0 = L_0^2(\Omega)$ such that

$$\tilde{a}(\mathbf{u}, \mathbf{u}, \mathbf{v}) + c(\mathbf{u}, \mathbf{u}, \mathbf{v}) + b(\mathbf{v}, p) = (\mathbf{f}, \mathbf{v}) \qquad \text{for all } \mathbf{v} \in \mathbf{V}_0 \quad (17.7)$$

and

$$b(\mathbf{u}, q) = 0 \qquad \text{for all } q \in S_0, \qquad (17.8)$$

where the forms $b(\cdot, \cdot)$ and $c(\cdot, \cdot, \cdot)$ are defined by (1.7) and (1.8), respectively, and where

$$\tilde{a}(\mathbf{w}, \mathbf{u}, \mathbf{v}) = \frac{1}{2} \int_\Omega \mathcal{C}(\mathbf{w})(\text{grad } \mathbf{u} + (\text{grad } \mathbf{u})^T) : (\text{grad } \mathbf{v} + (\text{grad } \mathbf{v})^T) \, d\Omega.$$

Results concerning the existence, uniqueness, and regularity of solutions of (17.7)–(17.8) may be found in Du and Gunzburger [1989a] and Ladyzhenskaya [1969, 1970a, and 1970b]. Also, through appropriate integration by parts procedures, it is easily shown that sufficiently smooth solutions of (17.7)–(17.8) satisfy (17.2)–(17.5).

Finite element discretizations of (17.7)–(17.8) may be defined in the usual manner. One chooses finite dimensional subspaces $\mathbf{V}_0^h \subset \mathbf{V}_0$ and $S_0^h \subset S_0$ and then seeks $\mathbf{u}^h \in \mathbf{V}_0^h$ and $p^h \in S_0^h$ such that

$$\tilde{a}(\mathbf{u}^h, \mathbf{u}^h, \mathbf{v}^h) + c(\mathbf{u}^h, \mathbf{u}^h, \mathbf{v}^h) + b(\mathbf{v}^h, p^h) = (\mathbf{f}, \mathbf{v}^h) \qquad \text{for all } \mathbf{v}^h \in \mathbf{V}_0^h$$

$$\tag{17.9}$$

and

$$b(\mathbf{u}^h, q^h) = 0 \qquad \text{for all } q^h \in S_0^h. \tag{17.10}$$

The discrete problem (17.9)–(17.10) is studied in Du and Gunzburger [1989b], which we follow below.

Some natural questions one may ask are concerned with whether finite element spaces that may be used for Navier–Stokes calculations may also be used in the discrete problem (17.9)–(17.10). For example, are finite element velocity and pressure spaces that are div-stable in the Navier–Stokes setting also div-stable for the Ladyzhenskaya equations? This is not altogether a trivial question because of the different underlying velocity spaces employed, i.e., $\mathbf{H}^1(\Omega) = \mathbf{W}^{1,2}(\Omega)$ versus $\mathbf{W}^{1,r+2}(\Omega)$ with $r > 0$. Then one may ask if the accuracy, measured in the same norms, obtainable through the use of a specific choice of finite element spaces is the same in the Ladyzhenskaya setting as it is for the Navier–Stokes equations. Both these questions can be answered in the affirmative.

Another issue is concerned with the use of quadrature rules in defining the discrete equations that are actually used in calculations. In Chapter 4 we saw that one need only integrate the linear viscous term in the Navier–Stokes equations exactly in order to achieve the approximation theoretic rates of convergence; the convection term need not be integrated exactly. In (17.9)–(17.10) there are additional nonlinearities present due to the modified constitutive law being used. However, it is known, for the case $r = 2$, that the same quadrature rules that preserve the optimal rates of convergence of the approximations in the Navier–Stokes case do likewise in the present case.

The local quadratic convergence of Newton's method for the solution of (17.9)–(17.10) can also be proven. For the case $r = 2$, Newton's method is defined as follows. Let

$$\hat{a}(\mathbf{w}, \mathbf{u}, \mathbf{v})$$

$$= \int_{\Omega} \nu_1(\text{grad } \mathbf{w} : \text{grad } \mathbf{u})(\text{grad } \mathbf{w} + (\text{grad } \mathbf{w})^T) : (\text{grad } \mathbf{v} + (\text{grad } \mathbf{v})^T) \, d\Omega$$

so that, for example,

$$\tilde{a}(\mathbf{u}, \mathbf{u}, \mathbf{v})$$

$$= \frac{1}{2} \hat{a}(\mathbf{u}, \mathbf{u}, \mathbf{v}) + \frac{1}{2} \nu \int_{\Omega} [\text{grad } \mathbf{u} + (\text{grad } \mathbf{u})^T] : [\text{grad } \mathbf{v} + (\text{grad } \mathbf{v})^T] \, d\Omega.$$

Then, given an initial guess $\mathbf{u}^{(0)}$ for the discrete velocity $\mathbf{u}^h$, the sequence of Newton iterates $\{\mathbf{u}^{(k)}, p^{(k)}\}$, $k \geq 1$, is found by solving the linear systems

$$\tilde{a}(\mathbf{u}^{(k-1)}, \mathbf{u}^{(k)}, \mathbf{v}^h) + c(\mathbf{u}^{(k-1)}, \mathbf{u}^{(k)}, \mathbf{v}^h) + c(\mathbf{u}^{(k)}, \mathbf{u}^{(k-1)}, \mathbf{v}^h)$$

$$+ b(\mathbf{v}^h, p^{(k)}) + \hat{a}(\mathbf{u}^{(k-1)}, \mathbf{u}^{(k)}, \mathbf{v}^h) = (\mathbf{f}, \mathbf{v}^h)$$

$$+ c(\mathbf{u}^{(k-1)}, \mathbf{u}^{(k-1)}, \mathbf{v}^h) + \hat{a}(\mathbf{u}^{(k-1)}, \mathbf{u}^{(k-1)}, \mathbf{v}^h) \qquad \text{for all } \mathbf{v}^h \in \mathbf{V}_0^h$$

and

$$b(\mathbf{u}^{(k)}, q^h) = 0 \qquad \text{for all } q^h \in S_0^h.$$

The results discussed above show that only a few changes need be made to an existing Navier–Stokes code in order to convert it into a program for the computation of approximations to the Ladyzhenskaya equations (17.2)–(17.4). One may use the same finite element spaces, e.g., the same basis functions defined with respect to the same grid, and the same quadrature rule. Indeed, one need only change that portion of the code that defines the entries of the discrete equations. This usually requires amending only a few lines of code.

An alternative for approximating solutions of (17.9)–(17.10) is to uncouple the velocity–pressure computation from the nonlinear viscosity. For example, instead of (17.9)–(17.10), one may discretize the sequence of problems

$$\tilde{a}(\mathbf{u}^{(m-1)}, \mathbf{u}^{(m)}, \mathbf{v}^h) + c(\mathbf{u}^{(m)}, \mathbf{u}^{(m)}, \mathbf{v}^h) + b(\mathbf{v}^h, p^{(m)}) = (\mathbf{f}, \mathbf{v}^h)$$

$$\text{for all } \mathbf{v}^h \in \mathbf{V}_0^h$$

and

$$b(\mathbf{u}^{(m)}, q^h) = 0 \qquad \text{for all } q^h \in S_0^h,$$

starting with some initial velocity $\mathbf{u}^{(0)}$. For $r = 2$ and with

$$a_t(\mathbf{u}, \mathbf{v}; v_t) = \frac{1}{2} \int_\Omega v_t(\mathrm{grad}\ \mathbf{u} + (\mathrm{grad}\ \mathbf{u})^T) : (\mathrm{grad}\ \mathbf{v} + (\mathrm{grad}\ \mathbf{v})^T)\, d\Omega$$

and

$$v_t(\mathbf{u}) = \mathcal{Q}(\mathbf{u}) = v + v_1 \left[\sum_{i,j=1}^{n} \left(\frac{\partial u_i}{\partial x_j} \right)^2 \right],$$

the above iteration is equivalent to, given a starting value for $v_t^{(0)}$, defining the sequence $\{\mathbf{u}^{(m)}, p^{(m)}, v_t^{(m)}\}$, for $m \geq 1$, to be the solutions of the sequence of problems

$$a_t(\mathbf{u}^{(m)}, \mathbf{v}; v_t^{(m-1)}) + c(\mathbf{u}^{(m)}, \mathbf{u}^{(m)}, \mathbf{v}) + b(\mathbf{v}, p^{(m)}) = (\mathbf{f}, \mathbf{v}) \qquad \text{for all } \mathbf{v} \in V_0,$$
$$(17.11)$$

$$b(\mathbf{u}^{(m)}, q) = 0 \qquad \text{for all } q \in S_0,$$
$$(17.12)$$

and

$$v_t^{(m)} = \mathcal{Q}(\mathbf{u}^{(m)}).\qquad (17.13)$$

Of course, (17.11)–(17.13), when discretized, will yield a nonlinear system of equations because we have not linearized the convection term. However, (17.11)–(17.13) are exactly a weak formulation of the Navier–Stokes equations with an effective, nonconstant viscosity given by $v_t^{(m-1)}$. The latter is evaluated by (17.13), which is simply an algebraic relation involving the derivatives of the components of the previous guess for the velocity. Thus, for each m, one may solve for $\mathbf{u}^{(m)}$ and $p^{(m)}$ using a (variable viscosity) Navier–Stokes code, with the only change being the trivial addition of code necessary to compute $v_t^{(m)}$ from $\mathbf{u}^{(m)}$.

The iteration (17.11)–(17.13) and its finite element approximation are considered in Gunzburger and Turner [1988] and in Turner [1989]. The important result is that $\mathbf{u}^{(m)} \to \mathbf{u}$ and $p^{(m)} \to p$ as $m \to \infty$, where $\mathbf{u}$ and p denote the solution of (17.7)–(17.8).

One may replace $\mathcal{Q}(\cdot)$ in (17.5) by

$$\mathcal{Q}(\mathbf{u}) = v \left[1 + \lambda \sum_{i,j=1}^{n} \left(\frac{\partial u_i}{\partial x_j} \right)^2 \right]^{(r-1)/2},\qquad (17.14)$$

where $v > 0$, $\lambda > 0$, and $0 < r < 1$ are constants. This relation models certain types of simple flows of polymers and polymer solutions and is known as the Carreau model. Note that since $(r - 1) < 0$, unlike the Ladyzhenskaya or algebraic turbulence models, high shear rates now imply a reduction in the effective viscosity. However, like those other models, for low shear rates the Navier–Stokes equations are recovered.

From an algorithmic point of view, there is no difficulty replacing (17.5) with (17.14). However, no analyses have been carried out concerning the stability and accuracy of finite element approximations of the solution of (17.2)–(17.4) with $\alpha(\cdot)$ defined by (17.14).

18

Bingham Fluids

18.1. Variational Inequalities Associated with Bingham Fluids

We wish to study the motion of a Bingham fluid in a bounded open subset Ω of $\mathbb{R}^n$, $n = 2$ or 3. A Bingham fluid is an example of a visco-plastic fluid. Typically, these fluids have the characteristic that the stress must be sufficiently large, or more precisely, some functional related to the stress must be sufficiently large, in order for the material to behave like a fluid. Otherwise, the material behaves like a rigid medium. The usual conservation laws are assumed to hold, and therefore these features of the fluid are wholly determined by its defining constitutive relation. For detailed discussions concerning Bingham fluids and mathematical analyses associated with their governing equations, one should consult Duvaut and Lions [1976] and Kim [1987 and 1989a]. From a numerical point of view, see Begis [1972]; Begis and Glowinski [1982]; Fortin [1972 and 1976]; Glowinski [1984]; Glowinski, Lions, and Trémolières [1981]; and Kim [1989b].

If $\mathbf{u}$ denotes the velocity field, we define

$$D_{ij}(\mathbf{u}) = \frac{1}{2}\left(\frac{\partial u_i}{\partial x_j} + \frac{\partial u_j}{\partial x_i}\right)$$

and

$$D_{\mathrm{II}}(\mathbf{u}) = \frac{1}{2} \sum_{i,j=1}^{n} (D_{ij}(\mathbf{u}))^2;$$

then the constitutive law for a Bingham fluid is given by

$$\sigma_{ij} = -p\delta_{ij} + \frac{gD_{ij}}{(D_{\mathrm{II}})^{1/2}} + 2\mu D_{ij}, \tag{18.1}$$

whenever $D_{\mathrm{II}} \neq 0$, where σ_{ij} denotes the stress tensor. In (18.1) p may be identified with the pressure and is given by $p = -(\sigma_{11} + \sigma_{22} + \sigma_{33})/3$. Also, μ denotes the usual dynamic viscosity coefficient, and g denotes the yield stress. The origin of the latter terminology is transparent when one inverts (18.1). If we let $\sigma_{ij}^{D} = \sigma_{ij} + p\delta_{ij}$ and

$$\sigma_{\mathrm{II}}(\mathbf{u}) = \frac{1}{2} \sum_{i,j=1}^{n} [\sigma_{ij}^{D}(\mathbf{u})]^2,$$

we then have that

$$D_{ij} = 0 \quad \text{if } \sigma_{\mathrm{II}}^{1/2} < g \tag{18.2}$$

and

$$D_{ij} = \frac{1}{2\mu} \left(1 - \frac{g}{\sigma_{\mathrm{II}}^{1/2}}\right)\sigma_{ij}^{D} \quad \text{if } \sigma_{\mathrm{II}}^{1/2} \geq g. \tag{18.3}$$

Thus we see from the first of these that if $\sigma_{\mathrm{II}}^{1/2} < g$, the fluid behaves like a rigid body. See Duvaut and Lions [1976] for details.

In regions where (18.3) holds, we may describe the motion of the fluid in the usual way, using (18.1) for the stress-rate of strain relation. Thus, in these regions, we have, for stationary flows, that

$$\mathbf{u} \cdot \operatorname{grad} \mathbf{u} + \operatorname{grad} p - \operatorname{div}[gD_{ij}/(D_{\mathrm{II}})^{1/2} + 2\mu D_{ij}] = \mathbf{f} \tag{18.4}$$

and

$$\operatorname{div} \mathbf{u} = 0 \tag{18.5}$$

for $\mathbf{x} \in \Omega$ such that $\sigma_{\mathrm{II}}^{1/2} \geq g$. In regions where the material behaves like a rigid body, the motion is described by (18.2), i.e.,

$$D_{ij} = 0 \quad \text{for } \mathbf{x} \in \Omega \text{ such that } \sigma_{\mathrm{II}}^{1/2} < g. \tag{18.6}$$

To these relations we append the simple boundary condition

$$\mathbf{u} = \mathbf{0} \text{ on } \Gamma, \tag{18.7}$$

where Γ denotes the boundary of Ω.

Unfortunately, it is not known *a priori* where in Ω (18.4)–(18.5) hold and where (18.6) holds. There is an unknown free boundary between regions where these different systems defining the motion of the material are valid.

An effective means of treating such a free boundary problem is to recast the problem into a variational inequality. To this end, let

$$\tilde{a}(\mathbf{u}, \mathbf{v}) = 2\mu \int_\Omega \sum_{i,j=1}^n D_{ij}(\mathbf{u})D_{ij}(\mathbf{v}) \, d\Omega,$$

$$c(\mathbf{w}, \mathbf{u}, \mathbf{v}) = \int_\Omega \sum_{i,j=1}^n w_j \frac{\partial u_i}{\partial x_j} v_i \, d\Omega = \int_\Omega \mathbf{w} \cdot \text{grad } \mathbf{u} \cdot \mathbf{v} \, d\Omega,$$

and

$$J(\mathbf{v}) = 2g \int_\Omega D_{\mathrm{II}}(\mathbf{v})^{1/2} \, d\Omega.$$

Then, the problem (18.4)–(18.7) has the equivalent formulation (Duvaut and Lions [1976])

$$\tilde{a}(\mathbf{u}, \mathbf{v} - \mathbf{u}) + c(\mathbf{u}, \mathbf{u}, \mathbf{v}) + J(\mathbf{v}) - J(\mathbf{u}) \geq (\mathbf{f}, \mathbf{v} - \mathbf{u}) \qquad (18.8)$$

for all test functions $\mathbf{v}$ such that div $\mathbf{v} = 0$ in Ω and $\mathbf{v} = \mathbf{0}$ on Γ. Along with (18.8), the velocity field $\mathbf{u}$ is required to satisfy (18.5) and (18.7).

The fact that the functional $J(\cdot)$ is not differentiable is a serious problem in discretizing (18.8). For this reason we consider the regularized version of (18.8)

$$\tilde{a}(\mathbf{u}_\varepsilon, \mathbf{v} - \mathbf{u}_\varepsilon) + c(\mathbf{u}_\varepsilon, \mathbf{u}_\varepsilon, \mathbf{v}) + J_\varepsilon(\mathbf{v}) - J_\varepsilon(\mathbf{u}_\varepsilon) \geq (\mathbf{f}, \mathbf{v} - \mathbf{u}_\varepsilon), \qquad (18.9)$$

where

$$J_\varepsilon(\mathbf{v}) = 2g \int_\Omega [\varepsilon + D_{\mathrm{II}}(\mathbf{v})]^{1/2} \, d\Omega.$$

Of course, $\mathbf{u}_\varepsilon$ is also required to satisfy (18.5) and (18.7). It can be shown that $|\mathbf{u}_\varepsilon - \mathbf{u}|_1 \to 0$.

Since the functional $J_\varepsilon(\cdot)$ is differentiable, (18.9) has the equivalent form

$$\tilde{a}(\mathbf{u}_\varepsilon, \mathbf{v}) + c(\mathbf{u}_\varepsilon, \mathbf{u}_\varepsilon, \mathbf{v}) + (J_\varepsilon'(\mathbf{u}_\varepsilon), \mathbf{v}) = (\mathbf{f}, \mathbf{v}),$$

where

$$(J_\varepsilon'(\mathbf{u}), \mathbf{v}) = g \int_\Omega \sum_{i,j=1}^n \frac{D_{ij}(\mathbf{u})D_{ij}(\mathbf{v})}{(\varepsilon + D_{\mathrm{II}}(\mathbf{u}))^{1/2}} \, d\Omega$$

and where the test function $\mathbf{v}$ is required to satisfy div $\mathbf{v} = 0$ in Ω and $\mathbf{v} = \mathbf{0}$ on Γ.

18.2. Finite Element Approximations

Finite element methods for Bingham fluids are considered in the computational references cited above. Due to the lack of regularity ($\mathbf{u}$ merely belongs to $\mathbf{W}^{1,6}(\Omega)$ for $\Omega \subset \mathbb{R}^3$) of the solution of the Bingham flow problem, it is probably useless to use anything but piecewise linear elements for the velocity approximation, and, for the most part it is this discretization that is considered by the above authors. In most situations, it is allowable to neglect the convection term; however, one must deal with the incompressibility constraint (18.5).

For example, a discretization may be defined as follows. Choosing $\mathbf{V}_0^h$ to consist of piecewise linear velocities that vanish on the boundary of Ω, or more precisely, on the boundary of a polygonal approximation Ω_h to Ω, we then choose a corresponding pressure space $S_0^h \subset L_0^2(\Omega)$ such that the pair $\mathbf{V}_0^h$ and S_0^h satisfies the div-stability condition. We then seek $\mathbf{u}^h \in \mathbf{V}_0^h$ and $p^h \in S_0^h$ such that

$$\tilde{a}(\mathbf{u}^h, \mathbf{v}^h) + c(\mathbf{u}^h, \mathbf{u}^h, \mathbf{v}^h) + b(p^h, \mathbf{v}^h) + (J_\varepsilon'(\mathbf{u}^h), \mathbf{v}^h)$$

$$= (\mathbf{f}, \mathbf{v}^h) \quad \text{for all } \mathbf{v}^h \in \mathbf{V}_0^h \tag{18.10}$$

and

$$b(\mathbf{u}^h, q^h) = 0 \quad \text{for all } q^h \in S_0^h, \tag{18.11}$$

where

$$b(\mathbf{v}, q) = -\int_\Omega q \operatorname{div} \mathbf{v} \, d\Omega.$$

Estimates for the difference $\mathbf{u} - \mathbf{u}^h$ can be derived by separately estimating $\mathbf{u} - \mathbf{u}_\varepsilon$ and $\mathbf{u}^h - \mathbf{u}_\varepsilon$. Typical results, using piecewise linear elements for the velocity, are of the form

$$|\mathbf{u} - \mathbf{u}^h|_1 = O(\varepsilon^\alpha) + O(h^\beta).$$

In general, $0 < \alpha \leq 1$ and $0 < \beta \leq 1$. For example, for problems in two dimensions one can show that (Fortin [1972]; Glowinski [1984]; and Glowinski, Lions, and Trémolières [1981]) $\alpha = \beta = 1/2$, although it is not known if these are the optimal values for these exponents. However, almost certainly, in this case $\alpha < 1$ and $\beta < 1$. For three-dimensional settings, precise values of the above exponents may be gleaned from Kim [1989b]. In any case, it is evident that the achievable accuracy is very poor and, as was mentioned before, cannot be improved through the use of higher-order elements due to the lack of regularity of the solution.

In time-dependent settings, and again due to the lack of regularity, it is probably useless to discretize in time anything more sophisticated than the backward Euler scheme. For a discussion of time discretizations, see the references cited in the previous paragraph. Also, Glowinski [1984] and Glowinski, Lions, and Trémolières [1981] may be consulted for iterative methods for solving the discrete system of equations.

IX

ELECTROMAGNETICALLY OR THERMALLY COUPLED FLOWS

In this part we consider flows for which the influence of variations in the temperature or in the magnetic field appreciably affect the motion of the fluid. In each case we only consider one prototype setting, which serves as an illustration of coupled fluid systems and for which the analyses of finite element approximations exist in the literature.

19

Flows of Liquid Metals

19.1. A Coupled Navier-Stokes-Maxwell Equation Model

Magnetic fields can influence the motion of liquid metals and of ionized gases, i.e., plasmas. The latter type of flows are usually in the compressible regime, while the former may be safely modeled as an incompressible flow. Therefore, here we consider only the former.

The equations that govern the stationary flow of liquid metals are the equations of incompressible magnetohydrodynamics and are given by (Landau and Lifshitz [1987], Sermange and Temam [1983], and Shercliff [1965])

$$-\frac{1}{G^2} \Delta \mathbf{u} + \frac{1}{N} \mathbf{u} \cdot \operatorname{grad} \mathbf{u} + \operatorname{grad} p + \frac{1}{R_m} \mathbf{B} \times \operatorname{curl} \mathbf{B} = \mathbf{f} \text{ in } \Omega, \quad (19.1)$$

$$\operatorname{div} \mathbf{u} = 0 \text{ in } \Omega, \quad (19.2)$$

$$\Delta \mathbf{B} + R_m \operatorname{curl}(\mathbf{u} \times \mathbf{B}) = \mathbf{0} \text{ in } \Omega, \quad (19.3)$$

and

$$\operatorname{div} \mathbf{B} = 0 \text{ in } \Omega, \quad (19.4)$$

where Ω is a bounded, open subset of $\mathbb{R}^3$. The flow variables $\mathbf{u}$ and p retain their usual meanings of velocity and pressure, respectively, while $\mathbf{f}$ denotes the body force. The electromagnetic variable $\mathbf{B}$ denotes the magnetic field. If the flow and electrical variables are assumed to be suitably nondimensionalized, the constants G, N, and R_m, respectively, denote the Hartmann number, the interaction parameter, and the magnetic Reynolds number. For details concerning the derivation of (19.1)–(19.4) and the definition of the parameters, see one of the abovementioned references.

If $R_m \ll 1$, the magnetic field $\mathbf{B}$ is essentially unaffected by the motion of the fluid and may be computed independently of the flow variables, i.e., (19.3)–(19.4) require that $\Delta\mathbf{B} = \mathbf{0}$ and div $\mathbf{B} = 0$. Then the term involving $\mathbf{B}$ in (19.1) becomes a source term to be added to $\mathbf{f}$. Finite element methods for the case $R_m \ll 1$ are analyzed in Peterson [1988], where a slightly different formulation coupling the flow variables to the electric field is used.

On the boundary Γ of Ω we assume that

$$\mathbf{u} = \mathbf{0}, \tag{19.5}$$

$$\mathbf{B} \cdot \mathbf{n} = 0, \tag{19.6}$$

and

$$\text{curl } \mathbf{B} \times \mathbf{n} = R_m \mathbf{g} \times \mathbf{n}, \tag{19.7}$$

where the specified data $\mathbf{g}$ is the tangential component of the electric field $\mathbf{E}$. To see this we recall the electromagnetic relations

$$\mathbf{j} = \mathbf{E} + \mathbf{u} \times \mathbf{B} \qquad \text{and} \qquad \text{curl } \mathbf{B} = R_m \mathbf{j},$$

where $\mathbf{j}$ denotes the current density. When combined these yield

$$\text{curl } \mathbf{B} = R_m \mathbf{E} + R_m \mathbf{u} \times \mathbf{B}.$$

Now suppose the tangential component of the electric field is specified on Γ, i.e., $\mathbf{n} \times (\mathbf{E} \times \mathbf{n}) = \mathbf{E} - (\mathbf{E} \cdot \mathbf{n})\mathbf{n} = \mathbf{g}$ on Γ. Then, using (19.5), we have that $\mathbf{n} \times (\text{curl } \mathbf{B} \times \mathbf{n}) = R_m \mathbf{g}$ and (19.7) is recovered from the vector identity curl $\mathbf{B} \times \mathbf{n} = \mathbf{n} \times (\text{curl } \mathbf{B} \times \mathbf{n}) \times \mathbf{n}$. The boundary conditions (19.5)–(19.7) imply that the boundary Γ is a solid conducting wall. Finite element methods for the approximation of solutions of (19.1)–(19.7) are considered in, e.g., Gunzburger, Meir, and Peterson [1989], which we follow here. The case of $\mathbf{u} = \mathbf{u}_0 \neq \mathbf{0}$ on Γ is also considered in the cited reference.

19.2. Finite Element Approximations

Finite element approximations of the solution of (19.1)–(19.7) are based on a Galerkin weak formulation. In order to define the latter, we use, for the velocity and pressure, the familiar spaces $\mathbf{H}_0^1(\Omega)$ and $L_0^2(\Omega)$, respectively, defined in Chapter 1. For the magnetic field we introduce the space

$$\mathbf{H}_n^1(\Omega) = \{\mathbf{v} \in \mathbf{H}^1(\Omega) : \mathbf{v} \cdot \mathbf{n} = 0 \text{ on } \Gamma\}.$$

We slightly amend, from those given in Chapter 1, the definitions of the forms $a(\cdot, \cdot)$ and $c(\cdot, \cdot, \cdot)$ to account for the constants appearing in (19.1), i.e.,

$$\tilde{a}(\mathbf{u}, \mathbf{v}) = \frac{1}{G^2} \int_\Omega \text{grad } \mathbf{u} : \text{grad } \mathbf{v} \, d\Omega \qquad \text{for all } \mathbf{u}, \mathbf{v} \in \mathbf{H}_0^1(\Omega)$$

and

$$\tilde{c}(\mathbf{w}, \mathbf{u}, \mathbf{v}) = \frac{1}{N} \int_\Omega \mathbf{w} \cdot \text{grad } \mathbf{u} \cdot \mathbf{v} \, d\Omega \qquad \text{for all } \mathbf{u}, \mathbf{v}, \mathbf{w} \in \mathbf{H}_0^1(\Omega).$$

The form $b(\cdot, \cdot)$ is still defined by

$$b(\mathbf{v}, q) = -\int_\Omega q \, \text{div } \mathbf{v} \, d\Omega \qquad \text{for all } \mathbf{v} \in \mathbf{H}_0^1(\Omega) \text{ and } q \in L_0^2(\Omega).$$

We introduce the new forms

$$d(\mathbf{C}, \mathbf{B}, \mathbf{v}) = \int_\Omega \text{curl } \mathbf{C} \times \mathbf{B} \cdot \mathbf{v} \, d\Omega$$

$$\text{for all } \mathbf{B}, \mathbf{C} \in \mathbf{H}_n^1(\Omega) \text{ and } \mathbf{v} \in \mathbf{H}_0^1(\Omega),$$

$$e(\mathbf{B}, \mathbf{C}) = \int_\Omega [(\text{div } \mathbf{B})(\text{div } \mathbf{C}) + (\text{curl } \mathbf{B}) \cdot (\text{curl } \mathbf{C})] \, d\Omega$$

$$\text{for all } \mathbf{B}, \mathbf{C} \in \mathbf{H}_n^1(\Omega),$$

and

$$\langle \mathbf{v}, \mathbf{C} \rangle = \int_\Gamma \mathbf{v} \cdot \mathbf{C} \, d\Gamma \qquad \text{for all } \mathbf{C} \in \mathbf{H}^{1/2}(\Gamma) \text{ and } \mathbf{v} \in \mathbf{H}^{-1/2}(\Gamma),$$

where the fractional trace space $\mathbf{H}^{1/2}(\Gamma)$ is defined in Chapter 1 and $\mathbf{H}^{-1/2}(\Gamma)$ is the dual space to $\mathbf{H}^{1/2}(\Gamma)$.

The Galerkin weak formulation of (19.1)–(19.7) that we use is defined as follows. Given $\mathbf{f} \in \mathbf{H}^{-1}(\Omega)$ and $\mathbf{g} \times \mathbf{n} \in \mathbf{H}^{-1/2}(\Gamma)$, seek $\mathbf{u} \in \mathbf{H}_0^1(\Omega)$,

$p \in L_0^2(\Omega)$, and $\mathbf{B} \in \mathbf{H}_n^1(\Omega)$ such that

$$\tilde{a}(\mathbf{u}, \mathbf{v}) + \tilde{c}(\mathbf{u}, \mathbf{u}, \mathbf{v}) + b(\mathbf{v}, p) - \frac{1}{R_m} d(\mathbf{B}, \mathbf{B}, \mathbf{v})$$

$$= (\mathbf{f}, \mathbf{v}) \quad \text{for all } \mathbf{v} \in \mathbf{H}_0^1(\Omega), \tag{19.8}$$

$$b(\mathbf{u}, q) = 0 \quad \text{for all } q \in L_0^2(\Omega), \tag{19.9}$$

and

$$e(\mathbf{B}, \mathbf{C}) + R_m d(\mathbf{C}, \mathbf{B}, \mathbf{u}) = R_m \langle \mathbf{g} \times \mathbf{n}, \mathbf{C} \rangle \quad \text{for all } \mathbf{C} \in \mathbf{H}_n^1(\Omega). \tag{19.10}$$

At first glance it might appear that we are not requiring that (19.4) be satisfied, especially since the similar condition (19.2) is enforced, at least in a weak sense, through (19.9). However, proceeding formally, we see that integrating by parts in (19.10) yields, using the essential boundary condition $\mathbf{C} \cdot \mathbf{n} = 0$,

$$\int_\Omega [\mathrm{curl}(\mathrm{curl}\,\mathbf{B} - R_m \mathbf{u} \times \mathbf{B}) - \mathrm{grad}(\mathrm{div}\,\mathbf{B})] \cdot \mathbf{C}\, d\Omega$$

$$+ \int_\Gamma [\mathrm{curl}\,\mathbf{B} \times \mathbf{n} - R_m \mathbf{g} \times \mathbf{n}] \cdot \mathbf{C}\, d\Gamma = 0.$$

Choosing $\mathbf{C}$ to be a general irrotational vector that vanishes on Γ then yields that $\mathrm{grad}(\mathrm{div}\,\mathbf{B}) = 0$ so that $\mathrm{div}\,\mathbf{B} = \mathrm{constant}$. However, this constant vanishes since, of course, the essential boundary condition $\mathbf{B} \cdot \mathbf{n} = 0$ on Γ implies that $\mathrm{div}\,\mathbf{B}$ has zero mean over Ω. The same procedure doesn't work with the velocity since

$$\frac{1}{N} \mathbf{u} \cdot \mathrm{grad}\,\mathbf{u} + \frac{1}{R_m} \mathbf{B} \times \mathrm{curl}\,\mathbf{B} - \mathbf{f} \tag{19.11}$$

is, in general, not solenoidal. Indeed, the role of the $\mathrm{grad}\,p$ term in (19.1) is to remove the irrotational part of (19.11) and thus render the sum of (19.11) and $\mathrm{grad}\,p$ solenoidal.

Finite element discretizations are defined by choosing subspaces $\mathbf{V}_0^h \subset \mathbf{H}_0^1(\Omega)$, $S_0^h \subset L_0^2(\Omega)$, and $\mathbf{V}_n^h \subset \mathbf{H}_n^1(\Omega)$ and then seeking $\mathbf{u}^h \in \mathbf{V}_0^h$, $p^h \in S_0^h$, and $\mathbf{B}^h \in \mathbf{V}_n^h$ such that

$$\tilde{a}(\mathbf{u}^h, \mathbf{v}^h) + \tilde{c}(\mathbf{u}^h, \mathbf{u}^h, \mathbf{v}^h) + b(\mathbf{v}^h, p^h) - \frac{1}{R_m} d(\mathbf{B}^h, \mathbf{B}^h, \mathbf{v}^h)$$

$$= (\mathbf{f}, \mathbf{v}^h) \quad \text{for all } \mathbf{v}^h \in \mathbf{V}_0^h, \tag{19.12}$$

$$b(\mathbf{u}^h, q^h) = 0 \quad \text{for all } q^h \in S_0^h, \tag{19.13}$$

and

$$e(\mathbf{B}^h, \mathbf{C}^h) + R_m d(\mathbf{C}^h, \mathbf{B}^h, \mathbf{u}^h) = R_m \langle \mathbf{g} \times \mathbf{n}, \mathbf{C}^h \rangle \qquad \text{for all } \mathbf{C}^h \in \mathbf{V}_n^h.$$
(19.14)

The velocity and pressure spaces may be chosen as in the case of the Navier–Stokes equations. To achieve balanced approximations, the approximating space for the magnetic field should be chosen the same as that used for the velocity, except, of course, that the former is constrained to satisfy (19.6) and not (19.5). With such a choice, optimal error estimates can be derived whenever the velocity–pressure spaces satisfy the div-stability condition. For example, suppose the Taylor–Hood element pair is used for the velocity and pressure approximations. Then choose $\mathbf{V}_n^h$ to consist of vector valued functions, each of whose components are piecewise quadratic polynomials, with respect to the same grid as that used to define the Taylor–Hood spaces. In addition, the members of $\mathbf{V}_n^h$ are required to be continuous over Ω and to have vanishing normal components on Γ. Then should the solution of (19.8)–(19.10) be sufficiently smooth, i.e., $\mathbf{u} \in \mathbf{H}^3(\Omega) \cap \mathbf{H}_0^1(\Omega)$, $p \in \mathbf{H}^2(\Omega) \cap L_0^2(\Omega)$, and $\mathbf{B} \in \mathbf{H}^3(\Omega) \cap \mathbf{H}_n^1(\Omega)$,

$$|\mathbf{u} - \mathbf{u}^h|_1 + \|p - p^h\|_0 + |\mathbf{B} - \mathbf{B}^h|_1 = O(h^2).$$

The usual duality arguments can be used to derive an $O(h^3)$ error estimate for $\|\mathbf{u} - \mathbf{u}^h\|_0$ and $\|\mathbf{B} - \mathbf{B}^h\|_0$.

19.3. Iterative Solution Techniques

The solution of (19.12)–(19.14) may be effected by, e.g., Newton's method. Here we start with initial guesses $\mathbf{u}^{(0)} \in \mathbf{V}_0^h$ and $\mathbf{B}^{(0)} \in \mathbf{V}_n^h$ satisfying the essential boundary conditions and then determine the sequence $\{\mathbf{u}^{(k)} \in \mathbf{V}_0^h, p^{(k)} \in S_0^h, \mathbf{B}^{(k)} \in \mathbf{V}_n^h\}$ from the sequence of linear systems

$$\tilde{a}(\mathbf{u}^{(k)}, \mathbf{v}^h) + \tilde{c}(\mathbf{u}^{(k)}, \mathbf{u}^{(k-1)}, \mathbf{v}^h) + \tilde{c}(\mathbf{u}^{(k-1)}, \mathbf{u}^{(k)}, \mathbf{v}^h) + b(\mathbf{v}^h, p^{(k)})$$

$$- \frac{1}{R_m} d(\mathbf{B}^{(k)}, \mathbf{B}^{(k-1)}, \mathbf{v}^h) - \frac{1}{R_m} d(\mathbf{B}^{(k-1)}, \mathbf{B}^{(k)}, \mathbf{v}^h) = (\mathbf{f}, \mathbf{v}^h)$$

$$+ \tilde{c}(\mathbf{u}^{(k-1)}, \mathbf{u}^{(k-1)}, \mathbf{v}^h) - \frac{1}{R_m} d(\mathbf{B}^{(k-1)}, \mathbf{B}^{(k-1)}, \mathbf{v}^h) \qquad \text{for all } \mathbf{v}^h \in \mathbf{V}_0^h,$$
(19.15)

$$b(\mathbf{u}^{(k)}, q^h) = 0 \qquad \text{for all } q^h \in S_0^h,$$
(19.16)

and

$$e(\mathbf{B}^{(k)}, \mathbf{C}^h) + R_m d(\mathbf{C}^h, \mathbf{B}^{(k-1)}, \mathbf{u}^{(k)}) + R_m d(\mathbf{C}^h, \mathbf{B}^{(k)}, \mathbf{u}^{(k-1)})$$

$$= R_m \langle \mathbf{g} \times \mathbf{n}, \mathbf{C}^h \rangle + R_m d(\mathbf{C}^h, \mathbf{B}^{(k-1)}, \mathbf{u}^{(k-1)}) \quad \text{for all } \mathbf{C}^h \in \mathbf{V}_n^h. \tag{19.17}$$

It can be shown, at least in the case when (19.12)–(19.14) have a unique solution, that the Newton iterates converge to the solution of (19.12)–(19.14) locally and quadratically. The fact that the Newton iteration (19.15)–(19.17) is only locally convergent requires that the initial guess be sufficiently close to the solution of (19.12)–(19.14). This requires the use of, e.g., continuation methods, to generate good initial guesses.

On the other hand, at least in the case when (19.12)–(19.14) has a unique solution, the following iterative scheme can be shown to be globally, although only linearly, convergent. We start with initial guesses $\mathbf{u}^{(0)}$ and $\mathbf{B}^{(0)}$, which need not satisfy any boundary conditions or (19.13), and then determine the sequence $\{\mathbf{u}^{(k)} \in \mathbf{V}_0^h, \, p^{(k)} \in S_0^h, \, \mathbf{B}^{(k)} \in \mathbf{V}_n^h\}$ for $k \geq 1$ from the sequence of linear systems

$$\tilde{a}(\mathbf{u}^{(k)}, \mathbf{v}^h) + \tilde{c}(\mathbf{u}^{(k-1)}, \mathbf{u}^{(k)}, \mathbf{v}^h) + b(\mathbf{v}^h, p^{(k)})$$

$$- \frac{1}{R_m} d(\mathbf{B}^{(k)}, \mathbf{B}^{(k-1)}, \mathbf{v}^h) = (\mathbf{f}, \mathbf{v}^h) \quad \text{for all } \mathbf{v}^h \in \mathbf{V}_0^h, \tag{19.18}$$

$$b(\mathbf{u}^{(k)}, q^h) = 0 \quad \text{for all } q^h \in S_0^h, \tag{19.19}$$

and

$$e(\mathbf{B}^{(k)}, \mathbf{C}^h) + R_m d(\mathbf{C}^h, \mathbf{B}^{(k-1)}, \mathbf{u}^{(k)})$$

$$= R_m \langle \mathbf{g} \times \mathbf{n}, \mathbf{C}^h \rangle \quad \text{for all } \mathbf{C}^h \in \mathbf{V}_n^h. \tag{19.20}$$

Both iterative schemes (19.15)–(19.17) and (19.18)–(19.20) require the simultaneous solution of all the variables $\mathbf{u}^{(k)}$, $p^{(k)}$, and $\mathbf{B}^{(k)}$. Even in two-dimensional problems, where these constitute five scalar fields, the task of solving the linear systems (19.15)–(19.17) or (19.18)–(19.20) is formidable. Three more attractive iterative schemes, from this point of view, are given below. In these schemes, the computation of each velocity and pressure iterate uncouples from that of the corresponding magnetic field iterate. Unfortunately, no convergence analyses exist for any of the schemes. Certainly, they are at best linearly and locally convergent.

For the first scheme we start with initial guesses $\mathbf{u}^{(0)}$ and $\mathbf{B}^{(0)}$ and then solve the linear system

$$\tilde{a}(\mathbf{u}^{(k)}, \mathbf{v}^h) + \tilde{c}(\mathbf{u}^{(k-1)}, \mathbf{u}^{(k)}, \mathbf{v}^h) + b(\mathbf{v}^h, p^{(k)})$$

$$= \frac{1}{R_m} d(\mathbf{B}^{(k-1)}, \mathbf{B}^{(k-1)}, \mathbf{v}^h) + (\mathbf{f}, \mathbf{v}^h) \quad \text{for all } \mathbf{v}^h \in \mathbf{V}_0^h, \quad (19.21)$$

$$b(\mathbf{u}^{(k)}, q^h) = 0 \quad \text{for all } q^h \in S_0^h, \quad (19.22)$$

and

$$e(\mathbf{B}^{(k)}, \mathbf{C}^h) + R_m d(\mathbf{C}^h, \mathbf{B}^{(k)}, \mathbf{u}^{(k)})$$

$$= R_m \langle \mathbf{g} \times \mathbf{n}, \mathbf{C}^h \rangle \quad \text{for all } \mathbf{C}^h \in \mathbf{V}_n^h, \quad (19.23)$$

for $\{\mathbf{u}^{(k)} \in \mathbf{V}_0^h, p^{(k)} \in S_0^h, \mathbf{B}^{(k)} \in \mathbf{V}_n^h\}$, $k \geq 1$. We see that $\mathbf{u}^{(k)}$ and $p^{(k)}$ can be determined from (19.21)–(19.22) using the known old values of $\mathbf{u}^{(k-1)}$ and $\mathbf{B}^{(k-1)}$. Then, having obtained $\mathbf{u}^{(k)}$, (19.23) is used to determine $\mathbf{B}^{(k)}$. Note that (19.21)–(19.22) are equivalent to a linearization of the discrete Navier–Stokes equations with an effective source term $\mathbf{f} + (\text{curl } \mathbf{B}^{(k-1)} \times \mathbf{B}^{(k-1)})/R_m$. Thus, to solve (19.21)–(19.22) requires relatively few changes to an existing Navier–Stokes code using the same linearization technique.

A second uncoupled iterative method requires an initial guess for the velocity only. Then $\{\mathbf{u}^{(k)} \in \mathbf{V}_0^h, p^{(k)} \in S_0^h, \mathbf{B}^{(k)} \in \mathbf{V}_n^h\}$ for $k \geq 1$ are the solutions of the linear systems

$$e(\mathbf{B}^{(k)}, \mathbf{C}^h) + R_m d(\mathbf{C}^h, \mathbf{B}^{(k)}, \mathbf{u}^{(k-1)})$$

$$= R_m \langle \mathbf{g} \times \mathbf{n}, \mathbf{C}^h \rangle \quad \text{for all } \mathbf{C}^h \in \mathbf{V}_n^h, \quad (19.24)$$

$$\tilde{a}(\mathbf{u}^{(k)}, \mathbf{v}^h) + \tilde{c}(\mathbf{u}^{(k-1)}, \mathbf{u}^{(k)}, \mathbf{v}^h) + b(\mathbf{v}^h, p^{(k)})$$

$$- \frac{1}{R_m} d(\mathbf{B}^{(k)}, \mathbf{B}^{(k)}, \mathbf{v}^h) = (\mathbf{f}, \mathbf{v}^h) \quad \text{for all } \mathbf{v}^h \in \mathbf{V}_0^h, \quad (19.25)$$

and

$$b(\mathbf{u}^{(k)}, q^h) = 0 \quad \text{for all } q^h \in S_0^h. \quad (19.26)$$

Thus now we solve (19.24) for the magnetic field $\mathbf{B}^{(k)}$ using the known old velocity $\mathbf{u}^{(k-1)}$ and then, having obtained $\mathbf{B}^{(k)}$, we solve (19.25)–(19.26) for the velocity $\mathbf{u}^{(k)}$ and pressure $p^{(k)}$. Again, note that (19.25)–(19.26) are equivalent to a linearization of the discrete Navier–Stokes equations, now with an effective source term $\mathbf{f} + (\text{curl } \mathbf{B}^{(k)} \times \mathbf{B}^{(k)})/R_m$. Thus, again, to solve (19.25)–(19.26) requires relatively few changes to an existing Navier–Stokes code using the same linearization technique.

For the final scheme we start with initial guesses $\mathbf{u}^{(0)}$ and $\mathbf{B}^{(0)}$ and then solve the linear system

$$\tilde{a}(\mathbf{u}^{(k)}, \mathbf{v}^h) + \tilde{c}(\mathbf{u}^{(k-1)}, \mathbf{u}^{(k)}, \mathbf{v}^h) + b(\mathbf{v}^h, p^{(k)})$$

$$= \frac{1}{R_m} d(\mathbf{B}^{(k-1)}, \mathbf{B}^{(k-1)}, \mathbf{v}^h) + (\mathbf{f}, \mathbf{v}^h) \quad \text{for all } \mathbf{v}^h \in \mathbf{V}_0^h, \quad (19.27)$$

$$b(\mathbf{u}^{(k)}, q^h) = 0 \quad \text{for all } q^h \in S_0^h, \quad (19.28)$$

and

$$e(\mathbf{B}^{(k)}, \mathbf{C}^h) + R_m d(\mathbf{C}^h, \mathbf{B}^{(k)}, \mathbf{u}^{(k-1)})$$

$$= R_m \langle \mathbf{g} \times \mathbf{n}, \mathbf{C}^h \rangle \quad \text{for all } \mathbf{C}^h \in \mathbf{V}_n^h, \quad (19.29)$$

for $\{\mathbf{u}^{(k)} \in \mathbf{V}_0^h, p^{(k)} \in S_0^h, \mathbf{B}^{(k)} \in \mathbf{V}_n^h\}$, $k \geq 1$. We see that $\mathbf{u}^{(k)}$ and $p^{(k)}$ can be determined from (19.27)–(19.28) using the known old values of $\mathbf{u}^{(k-1)}$ and $\mathbf{B}^{(k-1)}$. Simultaneously, (19.29) can be used to determine $\mathbf{B}^{(k)}$ from the old value of $\mathbf{u}^{(k-1)}$. Thus, not only do the flow and magnetic calculations uncouple at each step of the iteration, but these may be performed in parallel.

20

The Boussinesq Equations

20.1. The Boussinesq Model for Thermal Conduction in Fluids

One of the major simplifications that the assumption of incompressibility enables is the uncoupling, in most situations, of thermal and dynamic phenomena. Thus, one is able to solve for the pressure and velocity without regard to the energy equation, and invoke the latter only if the temperature field is desired. However, there are some flows for which this uncoupling is not justified, e.g., free convection. Typically, in these flows the effects of gravity or buoyancy cannot be neglected. On the other hand, consistent with the incompressibility assumption, one is still able to invoke assumptions concerning thermal phenomena that effect substantial simplifications in the mathematical model. Here, we examine the *Boussinesq model* for thermally coupled incompressible flows.

The derivation of the Boussinesq equations (Landau and Lifshitz [1987]) is based on four assumptions about the thermodynamics and thermal effects of the flow. The first is that variations in the density are negligible, except for the body force term in the momentum equation, which is given by $\rho\mathbf{g}$, where ρ denotes the density and $\mathbf{g}$

denotes the constant acceleration due to gravity. Next, it is assumed that the density in the term $\rho \mathbf{g}$ is given by $\rho = \rho_0[1 - \beta(T - T_0)]$, where ρ_0 and T_0 are a reference density and temperature, respectively, T the absolute temperature, and β is the thermal expansion coefficient. Moreover, we assume that, in the energy equation, one may neglect the dissipation of mechanical energy and that the viscosity μ, thermal expansion β, thermal conductivity κ, and specific heat at constant pressure c_p are all constants. The net result of these assumptions, if we in addition assume that the flow is stationary, is

$$-\mu \operatorname{div}[\operatorname{grad} \mathbf{u} + (\operatorname{grad} \mathbf{u})^T] + \rho_0 \mathbf{u} \cdot \operatorname{grad} \mathbf{u} + \operatorname{grad} p$$
$$= \mathbf{g}\rho_0[1 - \beta(T - T_0)] \text{ in } \Omega, \tag{20.1}$$

$$\operatorname{div} \mathbf{u} = 0 \text{ in } \Omega, \tag{20.2}$$

and

$$-\kappa \Delta T + \rho_0 c_p \mathbf{u} \cdot \operatorname{grad} T = Q \text{ in } \Omega, \tag{20.3}$$

where Q denotes a heat source and Ω is a bounded open set in $\mathbb{R}^n$, $n = 2$ or 3.

We assume that there is a length scale ℓ and a temperature scale $T_1 - T_0$ inherent in the problem, e.g., the distance and temperature difference between two walls. Then we may define the nondimensional Prandtl number $Pr = \kappa/\mu c_p$ and Grashof number $Gr = \beta \ell^3 \rho_0^2 |\mathbf{g}| (T_1 - T_0)/\mu^2$. If, in addition, there is a velocity scale U inherent in the problem, we may define the Reynolds number $Re = \rho_0 U\ell/\mu$; if there is no such velocity scale, e.g., for free convection, we choose $U = \mu/(\rho_0 \ell)$ so that in this case we set $Re = 1$. Then, if we nondimensionalize according to $\mathbf{x} \leftarrow \mathbf{x}/\ell$, $\mathbf{u} \leftarrow \mathbf{u}/U$, $T \leftarrow (T - T_0)/(T_1 - T_0)$, and $p \leftarrow (p - \mathbf{g} \cdot \mathbf{x})/(\rho_0 U^2)$, (20.1)–(20.3) yield that

$$-\frac{1}{Re} \operatorname{div}[\operatorname{grad} \mathbf{u} + (\operatorname{grad} \mathbf{u})^T] + \mathbf{u} \cdot \operatorname{grad} \mathbf{u} + \operatorname{grad} p = \frac{Gr}{Re^2} \mathbf{g} T \text{ in } \Omega, \tag{20.4}$$

$$\operatorname{div} \mathbf{u} = 0 \text{ in } \Omega, \tag{20.5}$$

and

$$-\frac{1}{PrRe} \Delta T + \mathbf{u} \cdot \operatorname{grad} T = Q \text{ in } \Omega. \tag{20.6}$$

In (20.4), $\mathbf{g}$ is now a unit vector in the direction of the gravitational acceleration and in (20.6), Q is now a nondimensional heat source.

One sees that a small Grashof number implies that the dynamic variables $\mathbf{u}$ and p are not influenced by temperature variations.

We again note that for free convection problems and, in general, for problems without an inherent velocity scale, one may set $Re = 1$.

A variety of boundary conditions are useful to describe practical flows. For example, if Γ denotes the boundary of Ω and Γ_i, $i = 1, \ldots, 4$, denote boundary segments such that $\Gamma_1 \cup \Gamma_2 = \Gamma$ and $\Gamma_3 \cup \Gamma_4 = \Gamma$, one may specify

$$\mathbf{u} = \mathbf{u}_d \text{ on } \Gamma_1, \tag{20.7}$$

$$\mathbf{u} \cdot \mathbf{n} = u_n \quad \text{and} \quad \mu\mathbf{n} \cdot [\text{grad } \mathbf{u} + (\text{grad } \mathbf{u})^T] \times \mathbf{n} = 0 \text{ on } \Gamma_2, \tag{20.8}$$

$$T = T_d \text{ on } \Gamma_3, \tag{20.9}$$

and

$$\mathbf{n} \cdot \text{grad } T = T_n \text{ on } \Gamma_4. \tag{20.10}$$

As is stands, (20.8) with $u_n = 0$ is the condition of free slip. Also, one may choose, for the temperature scaling, T_0 and T_1 to be the minimum and maximum values, respectively, of T_d, whenever the latter is not constant.

Another boundary condition of interest is one that accounts for effects due to surface tension. For example, if $\Omega \subset \mathbb{R}^2$ we may have, instead of (20.8),

$$\mathbf{u} \cdot \mathbf{n} = 0 \quad \text{and} \quad \mu\mathbf{n} \cdot [\text{grad } \mathbf{u} + (\text{grad } \mathbf{u})^T] \cdot \tau = -\alpha\mathbf{n} \cdot \text{grad } T \text{ on } \Gamma_2,$$

where τ denotes the unit tangential vector to Γ_2. This boundary condition arises, e.g., in crystal growth problems. For a discussion of more general boundary conditions than those given by (20.7)–(20.10) and finite element algorithms that can account for them, one may consult Driessen [1984].

Here we will consider only the case of homogeneous versions of (20.7)–(20.10), i.e., $\mathbf{u}_d = \mathbf{0}$, $u_n = T_d = T_n = 0$. The inhomogeneous case can be treated by the same type of techniques as were discussed in Chapter 4 for the case of the uncoupled Navier–Stokes equations.

20.2. Finite Element Approximations

As usual, in order to define a finite element algorithm, first we define a weak formulation of the problem. To this end we introduce the spaces

$$\mathbf{V}_0 = \{\mathbf{v} \in \mathbf{H}^1(\Omega) : \mathbf{v} = \mathbf{0} \text{ on } \Gamma_1 \quad \text{and} \quad \mathbf{v} \cdot \mathbf{n} = 0 \text{ on } \Gamma_2\},$$

$$S_0 = L_0^2(\Omega),$$

and

$$\Theta_0 = \{\theta \in H^1(\Omega) : \theta = 0 \text{ on } \Gamma_3\}.$$

We let Θ_0' denote the dual space of Θ_0. We also introduce the forms

$$\tilde{a}(\mathbf{u}, \mathbf{v}) = \frac{1}{2Re} \int_\Omega [\text{grad } \mathbf{u} + (\text{grad } \mathbf{u})^T] : [\text{grad } \mathbf{v} + (\text{grad } \mathbf{v})^T] \, d\Omega,$$

$$b(\mathbf{v}, q) = -\int_\Omega q \text{ div } \mathbf{v} \, d\Omega, \qquad c(\mathbf{w}, \mathbf{u}, \mathbf{v}) = \int_\Omega \mathbf{w} \cdot \text{grad } \mathbf{u} \cdot \mathbf{v} \, d\Omega,$$

$$d(T, \mathbf{v}) = -\frac{Gr}{Re^2} \int_\Omega T\mathbf{g} \cdot \mathbf{v} \, d\Omega,$$

$$e(T, \theta) = \frac{1}{PrRe} \int_\Omega \text{grad } T \cdot \text{grad } \theta \, d\Omega,$$

and

$$f(\mathbf{u}, T, \theta) = \int_\Omega \theta\mathbf{u} \cdot \text{grad } T \, d\Omega$$

defined for all $\mathbf{u}, \mathbf{v}, \mathbf{w} \in \mathbf{V}_0$, $q \in S_0$, and $T, \theta \in \Theta_0$.

A weak formulation of the problem (20.4)–(20.6) and homogeneous versions of (20.7)–(20.10) is then given as follows. Given $Q \in \Theta_0'$ and the unit vector $\mathbf{g}$, we seek $\mathbf{u} \in \mathbf{V}_0$, $p \in S_0$, and $T \in \Theta_0$ such that

$$\tilde{a}(\mathbf{u}, \mathbf{v}) + b(\mathbf{v}, p) + c(\mathbf{u}, \mathbf{u}, \mathbf{v}) + d(T, \mathbf{v}) = 0 \quad \text{for all } \mathbf{v} \in \mathbf{V}_0, \quad (20.11)$$

$$b(\mathbf{u}, q) = 0 \quad \text{for all } q \in S_0, \quad (20.12)$$

and

$$e(T, \theta) + f(\mathbf{u}, T, \theta) = (Q, \theta) \quad \text{for all } \theta \in \Theta_0. \quad (20.13)$$

Existence, uniqueness, and regularity results for solutions of (20.11)–(20.13) can be derived using the same techniques as that used for the uncoupled Navier–Stokes equations. For example, for free convection problems for which $Q = 0$ and $Re = 1$, the uniqueness of solutions can be proved only for sufficiently small Rayleigh number $Ra = GrPr$. See Lions [1969] and Cuvelier [1976 and 1978] for more information on these matters.

After choosing finite element subspaces $\mathbf{V}_0^h \subset \mathbf{V}_0$, $S_0^h \subset S_0$, and $\Theta_0^h \subset \Theta_0$, the discrete finite element equations may be defined as follows. We seek $\mathbf{u}^h \in \mathbf{V}_0^h$, $p^h \in S_0^h$, and $T^h \in \Theta_0^h$ such that

$$\tilde{a}(\mathbf{u}^h, \mathbf{v}^h) + b(\mathbf{v}^h, p^h) + c(\mathbf{u}^h, \mathbf{u}^h, \mathbf{v}^h) + d(T^h, \mathbf{v}^h) = 0 \quad \text{for all } \mathbf{v}^h \in \mathbf{V}_0^h,$$
$$(20.14)$$

$$b(\mathbf{u}^h, q^h) = 0 \quad \text{for all } q^h \in S_0^h,$$
$$(20.15)$$

and

$$e(T^h, \theta^h) + f(\mathbf{u}^h, T^h, \theta^h) = (Q, \theta^h) \quad \text{for all } \theta^h \in \Theta_0^h.$$
$$(20.16)$$

The finite element spaces for the velocity and pressure should be chosen as for the Navier–Stokes case; in particular, they should satisfy the div-stability condition. The finite element space for the temperature should be chosen to consist of piecewise polynomials of the same degree and with respect to the same triangulation as those used for the velocity components. Of course, these piecewise polynomials should be continuous over Ω and should satisfy the essential boundary condition $T^h = 0$ on Γ_3. With such a choice of approximating space, the error in the temperature approximation is comparable to that for the velocity.

As an example, consider using the Taylor–Hood element pair for the velocity and pressure. Then the temperature should be approximated with piecewise quadratic polynomials with respect to the triangulation used to define the Taylor–Hood spaces. Then, should the solution of (20.11)–(20.13) be sufficiently smooth, i.e., $\mathbf{u} \in \mathbf{H}^3(\Omega) \cap \mathbf{V}_0$, $p \in H^2(\Omega) \cap S_0$, and $T \in H^3(\Omega) \cap \Theta_0$, we have, at least for sufficiently small Grashof number,

$$|\mathbf{u} - \mathbf{u}^h|_1 + \|p - p^h\|_0 + |T - T^h|_1 = O(h^2).$$

The usual duality arguments can be used to derive an $O(h^3)$ error estimate for $\|\mathbf{u} - \mathbf{u}^h\|_0$ and $\|T - T^h\|_0$.

20.3. Iterative Solution Techniques

The discrete system (20.14)–(20.16) may be solved by a variety of methods. For example, Newton's method starts with initial guesses $\mathbf{u}^{(0)} \in \mathbf{V}_0^h$ and $T^{(0)} \in \Theta_0^h$, satisfying the essential boundary conditions, and then determines the sequence $\{\mathbf{u}^{(k)} \in \mathbf{V}_0^h, p^{(k)} \in S_0^h, T^{(k)} \in \Theta_0^h\}$ for

$k \geq 1$ to be the solution of the sequence of linear systems

$$\tilde{a}(\mathbf{u}^{(k)}, \mathbf{v}^h) + b(\mathbf{v}^h, p^{(k)}) + c(\mathbf{u}^{(k)}, \mathbf{u}^{(k-1)}, \mathbf{v}^h) + c(\mathbf{u}^{(k-1)}, \mathbf{u}^{(k)}, \mathbf{v}^h)$$

$$+ d(T^{(k)}, \mathbf{v}^h) = c(\mathbf{u}^{(k-1)}, \mathbf{u}^{(k-1)}, \mathbf{v}^h) \quad \text{for all } \mathbf{v}^h \in \mathbf{V}_0^h, \quad (20.17)$$

$$b(\mathbf{u}^{(k)}, q^h) = 0 \quad \text{for all } q^h \in S_0^h, \quad (20.18)$$

and

$$e(T^{(k)}, \theta^h) + f(\mathbf{u}^{(k)}, T^{(k-1)}, \theta^h) + f(\mathbf{u}^{(k-1)}, T^{(k)}, \theta^h)$$

$$= (Q, \theta^h) + f(\mathbf{u}^{(k-1)}, T^{(k-1)}, \theta^h) \quad \text{for all } \theta^h \in \Theta_0^h. \quad (20.19)$$

It can be shown that $\mathbf{u}^{(k)} \to \mathbf{u}^h$, $p^{(k)} \to p^h$, and $T^{(k)} \to T^h$ locally and quadratically (Dreissen [1984]), at least for cases where either or both T and $\mathbf{u} \cdot \mathbf{n}$ are specified at every point of the boundary.

As is always the case, the scheme (20.17)–(20.19) requires initial guesses $\mathbf{u}^{(0)}$ and $T^{(0)}$, which are sufficiently close to a solution $\mathbf{u}^h$ and T^h, respectively, of (20.14)–(20.16) in order for the iterates $\{\mathbf{u}^{(k)}, p^{(k)}, T^{(k)}\}$ to converge to $\{\mathbf{u}^h, p^h, T^h\}$. Thus, for high values of the Grashof number, and high values of the Reynolds number in cases with an inherent velocity scale, some sort of continuation method is necessary in order to generate a good initial guess for the solution at a desired value of these numbers.

For low values of Gr and Re, i.e., whenever the uniqueness of the solution of (20.11)–(20.13) can be guaranteed, the following iterative scheme can be shown to be globally and linearly convergent, at least for cases where either or both T and $\mathbf{u} \cdot \mathbf{n}$ are specified at every point of the boundary. We now start with an initial guess $\mathbf{u}^{(0)}$, which need not satisfy any boundary condition or (20.15), and then generate the sequence of iterates $\{\mathbf{u}^{(k)} \in \mathbf{V}_0^h, \ p^{(k)} \in S_0^h, \ T^{(k)} \in \Theta_0^h\}$ for $k \geq 1$ by solving the sequence of linear systems

$$e(T^{(k)}, \theta^h) + f(\mathbf{u}^{(k-1)}, T^{(k)}, \theta^h) = (Q, \theta^h) \quad \text{for all } \theta^h \in \Theta_0^h, \quad (20.20)$$

$$\tilde{a}(\mathbf{u}^{(k)}, \mathbf{v}^h) + b(\mathbf{v}^h, p^{(k)}) + c(\mathbf{u}^{(k-1)}, \mathbf{u}^{(k)}, \mathbf{v}^h)$$

$$+ d(T^{(k)}, \mathbf{v}^h) = 0 \quad \text{for all } \mathbf{v}^h \in \mathbf{V}_0^h, \quad (20.21)$$

and

$$b(\mathbf{u}^{(k)}, q^h) = 0 \quad \text{for all } q^h \in S_0^h. \quad (20.22)$$

Another advantage of (20.20)–(20.22) is that the temperature and the velocity–pressure calculations uncouple. In (20.17)–(20.19), the three variables $\mathbf{u}^{(k)}$, $p^{(k)}$, and $T^{(k)}$ are solved for simultaneously. On the other hand, using (20.20), one first solves for $T^{(k)}$ using the known old value of $\mathbf{u}^{(k-1)}$. Then having obtained $T^{(k)}$, one solves (20.21)–(20.22) for $\mathbf{u}^{(k)}$ and $p^{(k)}$.

Another uncoupled iterative scheme that evidently is only locally and linearly convergent (Dreissen [1984]) is given as follows. We start with initial guesses $\mathbf{u}^{(0)}$ and $T^{(0)}$ satisfying the same conditions as those for Newton's method. Then $\{\mathbf{u}^{(k)}, p^{(k)}, T^{(k)}\}$ for $k \geq 1$ are determined by solving the linear systems

$$\tilde{a}(\mathbf{u}^{(k)}, \mathbf{v}^h) + b(\mathbf{v}^h, p^{(k)}) + c(\mathbf{u}^{(k)}, \mathbf{u}^{(k-1)}, \mathbf{v}^h) + c(\mathbf{u}^{(k-1)}, \mathbf{u}^{(k)}, \mathbf{v}^h)$$

$$= -d(T^{(k-1)}, \mathbf{v}^h) + c(\mathbf{u}^{(k-1)}, \mathbf{u}^{(k-1)}, \mathbf{v}^h) \quad \text{for all } \mathbf{v}^h \in \mathbf{V}_0^h, \quad (20.23)$$

$$b(\mathbf{u}^{(k)}, q^h) = 0 \quad \text{for all } q^h \in S_0^h, \quad (20.24)$$

and

$$e(T^{(k)}, \theta^h) + f(\mathbf{u}^{(k)}, T^{(k)}, \theta^h) = (Q, \theta^h) \quad \text{for all } \theta^h \in \Theta_0^h. \quad (20.25)$$

Now, using (20.23)–(20.24) we solve for $\mathbf{u}^{(k)}$ and $p^{(k)}$ using the known old value of the temperature $T^{(k-1)}$. Then, having obtained $\mathbf{u}^{(k)}$, one uses (20.25) to obtain the temperature iterate $T^{(k)}$. Note that (20.23)–(20.24) are exactly the Newton equations for the uncoupled Navier-Stokes equations with an effective body force $(Gr\mathbf{g}T^{(k-1)}/Re^2)$. Thus, the solution of (20.23)–(20.24) may be effected through relatively few changes in an existing Navier–Stokes code. This is perhaps the only advantage of the scheme (20.23)–(20.25) over the scheme (20.20)–(20.22). Also note that (20.25) is simply a linear convection-diffusion equation for $T^{(k)}$. Of course, since in general (20.23)–(20.25) is only locally convergent, good initial guesses must be generated by, e.g., continuation methods.

X

REMARKS ON SOME TOPICS THAT HAVE NOT BEEN CONSIDERED

In this final chapter we consider, through very brief comments, a few of the important topics that we have not covered in the previous chapters. For each of these topics the cited references may be consulted for more complete treatments. It is especially noteworthy that we have hardly touched upon the two important issues of the numerical approximations of the many different models for turbulent flow and of fluid/structure interaction problems. Likewise, from an algorithmic viewpoint, we have not delved into two related methods, namely, finite volume and boundary integral methods.

21

Problems, Formulations, Algorithms, and Other Issues That Have Not Been Considered

Free Boundary Problems. An important class of problems in fluid mechanics is that of free boundary problems. Here, the boundary, or part of the boundary of the flow domain, or some interface within the domain, is unknown. To compensate for this lack of definition, one usually has available additional information, in the form of boundary conditions, along the unknown boundary. For example, one may impose both velocity and stress boundary conditions at the unknown portion of the boundary which, were that boundary actually known, would represent an overspecification of data for the problem.

One common method for solving free boundary problems is to guess the shape of the free boundary, imposing only a partial set of the given conditions that must hold there. Then, using the remaining conditions, one develops a new guess for the shape. One continues this iteration until convergence. The key to such iterations is to choose a method for updating the shape of the boundary. In design problems, where the boundary is to be determined subject to the extremization of some given functional, this is often a straightforward task. In other problems such as determining a water–air interface, developing an updating procedure requires substantial ingenuity. For details concerning finite element methods for free boundary problems associated with the

Navier–Stokes equations, one may consult Cuvelier, Segal, and van Steenhoven [1986]; Jean and Pritchard [1980]; Pironneau [1982]; and Tidd, Thatcher and Kaye [1986]. Also, recall that Bingham fluids provide another type of free boundary problem.

Iterative Solution Methods for Linear Systems. It is abundantly clear that problems posed on complex three-dimensional domains are not solvable, today or in the near future, if one uses direct methods for the solution of linear systems of algebraic equations. On the other hand, due to the lack of symmetry and/or positive definiteness of the linear systems arising from many formulation, discretization, and linearization combinations, successful iterative methods for the solution of linear systems have been hard to come by.

Two avenues suggest themselves, at least until the unlikely day that general purpose iterative linear system solvers are available. The first is to choose a formulation of the problem, a discretization scheme, and a linearization method for the nonlinear discrete equations that results in a sequence of symmetric, positive definite linear algebraic systems. Then these may be solved by, e.g., standard multigrid or preconditioned conjugate gradient methods. For example, see Ghia, Ghia, and Shin [1982] for an application of a multigrid method to an incompressible flow calculation. One motivation behind the rather great interest in using vorticity formulations or pressure Poisson equations (see below) is exactly that they have the promise that one may find approximate solutions through ultimately solving only symmetric, positive definite linear systems.

A second approach is to use the standard primitive variable formulation along with some discretization scheme to yield a nonlinear system of equations. One may then solve these equations by methods (see, e.g., Glowinski [1984], Temam [1979], or Thomasset[1981]) that require the solution of a sequence of discrete Stokes problems. The latter are linear and in general symmetric, but are not positive definite, so that standard iterative methods can not be directly applied. However, in Verfurth [1984b and 1988], multigrid methods for solving discrete Stokes problems are developed and analyzed. .

It should also be noted that many of the methods that ultimately yield linear systems that can be solved through iterative methods also seemingly yield to naive, and thus easily defined, parallel processing algorithms.

Thus, there is hope that iterative linear systems solvers may be of use in incompressible flow calculations. However, it should be pointed out that iterative methods for the nonlinear discrete equations that result in linear systems that are themselves amenable to solution through iterative methods are notoriously slow to converge, especially for high values of the Reynolds number. It is safe to say that the outstanding problem remaining connected with incompressible flow simulations is the development of efficient solution methods for the discrete equations.

Optimization and Control Problems.   Optimization and optimal control problems associated with incompressible viscous flows usually involve one or more control parameters and a functional depending on the flow variables and perhaps the control parameters. The goal is to minimize the given functional by judiciously choosing the control parameters.

Control may be effected in a variety of ways. One may control the flow through adjustments in the body force, or, more commonly, through the blowing or suction of fluid through orifices on the boundary. One can also control the flow by adjusting the shape of the flow domain.

As an example of a useful functional, consider

$$J(\mathbf{u}) = \int_{\Omega} [\text{grad } \mathbf{u} + (\text{grad } \mathbf{u})^T] : [\text{grad } \mathbf{u} + (\text{grad } \mathbf{u})^T]\, d\Omega,$$

which is proportional to the viscous drag on a body. Thus, given a fixed flow domain Ω with boundary Γ, we can ask for $(\mathbf{u}, p, \mathbf{g})$, which satisfies the Navier–Stokes equations (1.1) and (1.2) and the boundary condition $\mathbf{u} = \mathbf{g}$ on Γ and which minimizes $J(\cdot)$. Here $\mathbf{g}$ is the (boundary) control variable that is allowed to vary over some set of admissible controls. Alternately, we could ask for $(\mathbf{u}, p)$ satisfying (1.1)–(1.3) and a domain Ω such that $J(\cdot)$ is minimized. Here the domain Ω is allowed to vary over some prescribed class of domains. Time-dependent versions of these problems are also easily formulated.

Problems such as these have been analyzed, both with respect to properties of the solution of the continuous problem, e.g., existence and uniqueness, and with respect to finite element approximations. One may consult Pironneau [1984] for the shape control problem, and Gunzburger, Hou, and Svobodny [1989] for problems controlled by the velocity at the boundary. Other control problems for the Navier–Stokes equations are considered in Cuvelier [1976 and 1978], Fursikov [1982, 1983a and 1983b], and Lions [1985] and some of the references cited therein.

Outflow Boundary Conditions. There are few topics connected with the numerical simulation of incompressible flows that give rise to more opinions and disputes than does the choice of outflow conditions. This problem arises when one truncates a flow domain of infinite extent, thus introducing an artificial boundary. Points on this boundary for which $\mathbf{u} \cdot \mathbf{n} > 0$ form the outflow portion of the boundary, and at these points one must invoke an artificial boundary condition in order to close the problem. For exterior problems for the Stokes equations, we have already discussed some aspects of this problem in Part VII.

Since the solution is not known *a priori* at the outflow boundary, necessarily any local artificial boundary condition imposed there will be in error. One hopes to apply an outflow boundary condition that has a minimal effect on the flow upstream of the outflow boundary. For many practitioners, "minimal effect" simply means no artificially induced oscillations in the streamlines upstream of the outflow boundary.

It is important to note the effect of the incorrect outflow boundary condition on the solution of the continuous problem. There is evidence, see Fix and Gunzburger [1977] and Segal [1982], that the effects are restricted to a boundary layer adjacent to the outflow boundary, i.e., the flow adjusts to the wrong boundary condition over a thin region near the outflow boundary. This is not a boundary layer of the usual type, i.e., one in which the flow is mostly tangent to the boundary and of a thickness proportional to $Re^{-1/2}$. Instead, we have a boundary layer in which the flow is mostly normal to the boundary and is of a thickness proportional to Re^{-1}. Thus, even for moderate values of the Reynolds number, this outflow boundary layer is very thin. The implications on numerical simulations are as follows. If one is willing to resolve the outflow boundary layer, i.e., through mesh refinement near the outflow, then just about any outflow boundary condition can be used without causing upstream wiggles. In fact, upstream wiggles are an indication that, for the particular outflow boundary condition being employed, the outflow boundary layer is not being resolved. On the other hand, since the outflow boundary layer is thin and thus resolving it requires serious mesh refinement, one would rather choose an outflow boundary condition for which the boundary layer is not present. This is, of course, easier said than done and is the object of rather extensive attention and interest.

There are many sources that may be consulted for a discussion of outflow boundary conditions; indeed, just about any paper dealing with flows in regions of infinite extent has to deal with outflows. For example, see Cuvelier, Segal, and van Steenhoven [1986]; Gresho [1988]; Peyret and Taylor [1983]; Segal [1982]; and Roache [1972], just to name a few. There are, to this day, no agreed basic principles to guide one in the choice of outflow boundary conditions. There is even evidence that the nonphysical boundary condition (4.11.1) is useful at outflows; see Gresho [1988] and Glowinski [1984]. However, combinations of tangential velocity and normal stress boundary conditions seem to be gaining some popularity.

Another related idea, which has not been extensively explored in the context of problematic outflows, is to "parabolize" the Navier–Stokes equations in the vicinity of the outflow boundary. Thus, in that region, one drops the diffusion term in the direction normal to the outflow boundary, removing the necessity of applying a boundary condition there. For example, if the outflow boundary is perpendicular to the x_1 direction, then the $\nu\mathbf{u}_{x_1 x_1}$ term is dropped from (1.1) near the outflow boundary.

Pressure Poisson Equations. Taking the divergence of the momentum equation (1.1) yields, with the aid of the continuity equation (1.2),

$$\Delta p = \operatorname{div} \mathbf{f} - \operatorname{grad} \mathbf{u} : (\operatorname{grad} \mathbf{u})^T, \tag{21.1}$$

which has the appearance of a Poisson equation for the pressure. The same equation holds in the time-dependent case as well. It is often the case that this equation is substituted for the incompressibility constraint (1.2), i.e., one solves (1.1) and (21.1) for $\mathbf{u}$ and p. Two questions, which are not unrelated, arise immediately. First, what boundary conditions should be applied for the pressure along portions of the boundary where the velocity is specified? Second, if we solve (1.1) and (21.1), is the velocity field so obtained solenoidal? These questions and their answers have been extensively discussed in the literature. The most complete and up-to-date discussion may be found in Gresho and Sani [1988]. They conclude that the rational boundary condition to impose on the pressure is the normal component of the momentum equation, i.e., on portions of the boundary on which the velocity is specified we impose

$$\frac{\partial p}{\partial n} = \mathbf{n} \cdot (\mathbf{f} + \nu \, \Delta \mathbf{u} - \mathbf{u} \cdot \text{grad } \mathbf{u})$$

or, in the time-dependent case,

$$\frac{\partial p}{\partial n} = \mathbf{n} \cdot \left(\mathbf{f} - \frac{\partial \mathbf{u}}{\partial t} + \nu \, \Delta \mathbf{u} - \mathbf{u} \cdot \text{grad } \mathbf{u} \right).$$

In the latter case at least, it can be shown rigorously that this boundary condition will result in a solenoidal velocity field.

The attractions of using (21.1) are obvious. One avoids dealing with the incompressibility constraint (1.2); instead, one deals with a Poisson type equation that everyone would rather solve. Indeed, the way (1.1) and (21.1) are usually used is to guess a discrete velocity field and solve a discretized version of (21.1) for the pressure. Then, using the discrete pressure so obtained in (1.1), one proceeds to solve for a new discrete velocity field from some discretization of (1.1). These steps are repeated until (hopefully) convergence is achieved. One does not have to worry about choosing special finite element spaces for the velocity and pressure. The reference cited above may be consulted for more details concerning the approximation of the pressure Poisson equation.

It is also possible to formulate rational Dirichlet boundary conditions for the pressure. This is the essence of the "Glowinski–Pironneau scheme". For details, see Girault and Raviart [1986] and Glowinski and Pironneau [1979].

Spectral and p-Versions of the Finite Element Method. The p-version of the finite element method, and its close relative, spectral element methods, are alternatives to the h-version of the finite element method that we have been discussing all along. In the h-version, increased accuracy is achieved, using polynomials of fixed degree, by decreasing the size of the mesh. In the p-version, one keeps the mesh fixed and increases the degree of the polynomials employed. Of course, the two approaches may be combined into h–p-methods. Spectral element methods are essentially p-methods that are derived as generalizations of spectral methods, e.g., Fourier methods, to general domains. For a recent survey of the p- and h–p-versions of the finite element method, see Babuska [1988].

One key to the success of such methods is the availability of velocity and pressure spaces consisting of high-degree polynomials that also satisfy the div-stability condition. We have discussed this issue in Section 3.4. One potential difficulty with the p-method is that the constant γ appearing in the div-stability condition (2.10) goes to zero algebraically as p increases. In Jensen and Vogelius [1989] it is shown that in spite of this, and due to the good approximating capabilities of high-degree polynomials, promising error estimates can be obtained.

The actual use of these methods for viscous flow calculations has been largely confined to the spectral element approach. See Karniadakis, Bullister, and Patera [1986]; Korczak and Patera [1986]; and Patera [1984]. For an analysis of the approximation of solutions of the Stokes problem, see Jensen and Vogelius [1989] and Scott and Vogelius [1985b]. One observation that can be made from an examination of the latter two references is that div-stable high-order elements and associated local basis sets are derived from conforming, i.e., continuously differentiable, streamfunction fields and their local basis. Of course, one may use the latter directly in a streamfunction formulation of viscous incompressible flow; see Part V. Thus, it may be that p-methods will be used most often in this fourth-order setting.

Whether or not these methods are advantageous for viscous flow calculations has not yet been determined. Certainly the state of the art of the p- and h–p-versions is nowhere near as advanced in the fluid mechanics setting as it is for solid mechanics.

Streamfunction and/or Vorticity Formulations in Three Dimensions. For two-dimensional flow problems, of both the plane and axially symmetric type, the streamfunction–vorticity equations have been a very popular model for basing numerical simulations of viscous incompressible flows. It is natural to try to use a similar approach for three-dimensional problems. For example, one may still define a streamfunction, or, as it is sometimes referred to, a vector potential Ψ and then let $\mathbf{u} = \operatorname{curl} \Psi$. Thus the incompressibility condition div $\mathbf{u} = 0$ is again automatically satisfied. Then, for example, with $\omega = \operatorname{curl} \mathbf{u}$ we can define the three-dimensional streamfunction-vorticity equations

$$\operatorname{curl}(\operatorname{curl} \Psi) = \omega \quad \text{and} \quad \nu \operatorname{curl}(\operatorname{curl} \omega) + \operatorname{curl}(\omega \times \operatorname{curl} \Psi) = \operatorname{curl} \mathbf{f}.$$
$$(21.2)$$

Unfortunately, some of the advantages of the streamfunction–vorticity formulation for two-dimensional problems do not carry over to the three-dimensional case. First of all, we now have *more* unknown fields to solve for than in the primitive variable formulation, i.e., there are six scalar fields in ω and Ψ, while there are only four in u and p. Also, in two dimensions the streamfunction is uniquely determined up to an additive constant, while in three dimensions one may add the gradient of any scalar function to a candidate streamfunction and not effect any change in the velocity field. This implies that if one tries to solve the above equations for ω and Ψ one must somehow constrain Ψ. A popular choice is to require that div $\Psi = 0$. Of course, we would also like to have div $\omega = 0$. These may be used to transform (21.2) into

$$-\Delta\Psi = \omega \quad \text{and} \quad -\nu\,\Delta\omega + \text{curl}(\omega \times \text{curl}\,\Psi) = \text{curl}\,f. \quad (21.3)$$

The next problem that arises is the definition of boundary conditions for both ω and Ψ. This is a much tougher problem in three dimensions than it is in two dimensions. There are also problems encountered in attempting to discretize (21.2) or (21.3).

One could also eliminate the vorticity from the system (21.3) to obtain a single (vector valued) equation for the streamfunction or vector potential Ψ.

In spite of these problems, there has been some progress in attacking the three-dimensional streamfunction–vorticity and vector potential formulations, both from engineering and mathematical viewpoints. For example, see Bendali, Dominguez, and Gallic [1985]; Ruas [1986]; Sherif and Hafez [1983]; and Verfurth [1987b].

An alternative to the streamfunction–vorticity and vector potential formulations is the velocity–vorticity formulation. Here the basic equations are

$$\text{div}\,u = 0, \quad \text{curl}\,u = \omega, \quad \text{and} \quad -\nu\,\Delta\omega + \text{curl}(\omega \times u) = \text{curl}\,f.$$
$$(21.4)$$

These are sometimes replaced by

$$-\Delta u = \text{curl}\,\omega \quad \text{and} \quad -\nu\,\Delta\omega + \text{curl}(\omega \times u) = \text{curl}\,f. \quad (21.5)$$

These systems of equations have gained popularity, especially in the second case, mainly due to their appearance as six Poisson equations for the components of u and ω. However, the formulations (21.4) or (21.5) are not without their serious problems. For example, boundary conditions for the vorticity must be created at portions of the boundary where the velocity is specified. There are also problems with the accuracy of the discrete solution. See Guevremont *et al.* [1988]; Gunzburger, Mundt, and Peterson [1989]; and Osswald, Ghia, and Ghia [1988] and the references cited therein for discussions of the faults and virtues of the velocity–vorticity formulation and its numerical approximation.

Test Problems. Anytime one develops a numerical algorithm one would like to implement it in the form of a working code so that the algorithm's accuracy and efficiency can be demonstrated through computational experiments. In order to generate numbers, one must ultimately choose a particular physical problem for which the code provides a numerical simulation. There are a wide variety of such test problems being used. Here we discuss the relative merits of some of these.

Perhaps the most popular test problem is the driven cavity problem. Here the flow domain is a square (or cube in three dimensions). Along all walls except the top one, the velocity is required to vanish. Along the top wall the normal velocity component vanishes and the tangential components are prescribed constants. The Reynolds number for the flow, for a fluid of given dynamic viscosity, is determined by the size of the box and the magnitude of the nonvanishing velocity components at the boundary. A difficulty associated with the driven cavity problem is that the flow contains (at the corners) strong, non-physical singularities. The effect of these singularities is mitigated in various ways by smoothing out the transition between the boundary conditions at the corners. In spite of these difficulties, the driven cavity problem is too easy a test for candidate algorithms for the numerical simulation of viscous incompressible flows. The reason for this is that it is rather easy to generate solutions that have the global features one would expect in such a flow. Furthermore, since the problem is for the most part not physically realizable, one cannot compare the numerical solution with meaningful experimental data.

The second most popular test problem is the flow over a backward facing step. The flow configuration is sketched in Figure 21.1, where the left boundary is an inflow and the right boundary is an outflow. The bottom boundary is a solid wall, while the top boundary may be a wall or a boundary at which the flow field is essentially inviscid.

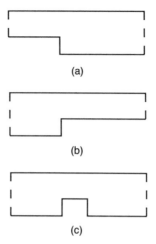

Figure 21.1

Three simple geometric configurations often used in test problems. (a) Backward facing step, (b) forward facing step, and (c) full step.

For the most part, the top boundary is chosen to be a solid wall. This problem is also too easy a test, mainly because certain benign features of the flow, e.g., the position of the reattachment point behind the step, scales with the Reynolds number; see Halim and Hafez [1984].

Perhaps the best test problems that retain the feature of geometrically simple flow domains are the forward-facing step and the full-step problems sketched in Figure 21.1. The distribution of boundary conditions is similar to that for the backward-facing step problem. These problems are realistic in the sense that they do not scale with the Reynolds number and that meaningful experimental data can be used for comparison purposes.

None of the problems discussed so far is solvable exactly and therefore one can ask how one should measure the quality of the numerical solution. Ideally, and where possible, one should compare with experimental data. However, this is not a foolproof measuring device since one seldom knows the accuracy of the experimental data itself. One popular method of determining the accuracy of a numerical solution is to use the "eyeball norm." A numerical simulation is good in the eyeball norm if one looks at a picture of the numerical flow field and concludes that it looks reasonable, e.g., there are no unexpected wiggles. Of course, this measure of the quality of a flow simulation, although popular, can be extremely misleading. For example, one can easily generate numerical flow fields whose gross features, e.g., recirculation regions, look reasonable but whose detailed features are wrong. A better way to judge the quality of the numerical solution is to compute using meshes of different sizes and show, using the type of norms introduced in our discussions, that convergence is apparent. Incidentally, the nonavailablity of exact solutions for most test problems points out the value of rigorous error estimates. After all, if we know *a priori* that the numerical solution converges to the (unknown) exact solution, and also know something about the rates of convergence, then we can have some confidence that we are producing meaningful numerical flow fields. Of course, the fly in the ointment is that error estimates are asymptotic in nature, i.e., they hold as the mesh size tends to zero, while actual computations are carried out using finite mesh sizes.

Ideally one would like to have exact solutions with which to compare one's numerical output. There are available many exact solutions of the Navier–Stokes equations, sometimes in the sense that they can be determined to arbitrary accuracy by solving nonlinear ordinary differential equations. Unfortunately, these solutions necessarily are ones that scale with the Reynolds number. Some, like fully developed Poiseuille flow, are exceedingly simple in that the nonlinear convection term vanishes. Others, such as Hiemenz and Hammel flow (Schlicting [1979]) are of more use. Although comparison with these solutions does not always provide definite information concerning the quality of a numerical simulation, a great amount of useful information can be so obtained. Typically, the flow domain for these solutions is unbounded, and exact solutions are obtained by solving nonlinear ordinary differential equations. A finite domain problem is obtained by truncating the flow domains. One is free to choose the shape of the artificial boundary so that, for example, one can also use these problems to test one's methods on flow domains having curved boundaries. Along the artificial boundaries created by the domain truncation process one can impose boundary conditions derived from the exact solution. However, one is free to choose any type of boundary condition one desires, e.g., the velocity, the stress, the vorticity, the pressure, or some combination of these. Thus, knowing an exact solution allows one to test how effective one's algorithms are in treating a variety of boundary conditions. In fact, along outflow portions of the artificial boundary one can test different (nonexact) outflow boundary conditions. Finally, the most obvious advantage of having an exact solution available is that one can precisely measure, using one's favorite norms, how good a numerical solution is.

Upwind, Petrov–Galerkin, and Streamwise Diffusion Methods.

For the most part, the schemes we have discussed in the book fall into the category of "central difference schemes" in the sense that nowhere in the discretization processes is there any bias towards any particular direction. Thus, the finite difference realizations of our schemes would involve only central difference quotient approximations to derivatives. It is well known that such approximations yield results that are often hopelessly polluted by oscillations whenever one does not adequately resolve regions where the flow variables experience large variations, e.g., boundary layers. At the outset of this discussion, it should be pointed out that if one does resolve such regions by, e.g., mesh refinement, then finite element methods as described in the bulk of the book will yield meaningful approximations.

However, mesh refinement for the purpose of keeping a calculation stable is often thought of as wasteful, i.e., one wants accuracy considerations to govern the selection of grids. For this reason, computational fluid dynamicists have long advanced the idea of upwind differencing in the convection term of the momentum equation, i.e., biasing, in the direction opposite to the flow, the differencing of that term. Crude methods for accomplishing this stabilize computations by introducing large amounts of artificial viscosity. This is fine in many settings, e.g., inviscid flows, where one is not interested in the details of the flow that are seriously influenced by the effects of viscosity. However, if one is really interested in viscous phenomena such as skin friction, then the introduction of large amounts of artificial viscosity can be ruinous.

The finite difference community has spent much time and effort in developing more sophisticated upwind schemes, where we use that term to denote any scheme that, either explicitly or implicitly, introduces a bias into the discretization. The finite element community has likewise been involved in such schemes, often drawing from techniques developed for finite difference methods, but sometimes developing new approaches that stem inherently from finite element methodology.

There are a variety of ways to introduce a bias into the discretization. One way is to choose test functions that differ from the basis functions for the trial space. This itself may be accomplished by a variety of means, perhaps the most rational of which is to test against a combination that looks like a linearized convection term. See Brooks and Hughes [1982], Christie *et al.* [1976], and Heinrich *et al.* [1977] for some specific examples of such "Petrov–Galerkin" formulations. These methods often suffer from the same problems as did crude upwind differencing techniques, namely, too much artificial viscosity, both in and orthogonal to the direction of the flow.

A more intriguing idea is that of streamwise upwinding. The motivation behind this idea is as follows. For a fixed mesh size, as one increases the Reynolds number Re, large, nonphysical oscillations can appear in the numerical flow field. As was mentioned already, these can be eliminated by upwinding/artificial diffusion techniques. However the very idea of letting the Reynolds number increase while keeping the mesh size fixed precludes the possibility of the accurate computation of viscous phenomena. For this one must at least let the mesh size go to zero as the Reynolds number increases. The precise dependence of h, the mesh size, on $1/Re$ depends on what viscous phenomena one is interested in simulating.

For example, the simulation of shear or tangential boundary layers along walls requires that $h = O(1/Re^{1/2})$ since that is roughly a measure of the thickness of those layers. Conceivably, especially at outflows, there may be other types of layers, perhaps induced numerically, requiring the more onerous restriction $h = O(1/Re)$. The latter layers may not be of physical interest and thus one is willing to smooth them out through the introduction of artificial diffusion. However, in so doing one does not want to smooth out the physically interesting tangential layers. Thus one is willing to pay the price of choosing $h = O(1/Re^{1/2})$, at least locally, in order to have an accurate resolution of, e.g., boundary layers along walls, but one would like to avoid paying the price of choosing $h = O(1/Re)$ in order to keep unwanted oscillations away. A method for accomplishing this is to use a tensor artificial viscosity, which adds damping in the streamwise direction, but none in directions perpendicular to the flow. One way to do this was introduced by Dukowicz and Ramshaw [1979]; they simply replace the viscous term $\nu \operatorname{div}[\operatorname{grad} \mathbf{u} + (\operatorname{grad} \mathbf{u})^T]$ in the momentum equation by

$$\operatorname{div}([\operatorname{grad} \mathbf{u} + (\operatorname{grad} \mathbf{u})^T](\nu I + \lambda \mathbf{u}\mathbf{u}^T))$$

for some parameter λ. Thus, we effectively have an anisotropic viscosity tensor $(\nu I + \lambda \mathbf{u}\mathbf{u}^T)$ where the added viscosity $(\lambda \mathbf{u}\mathbf{u}^T)$ acts only in the streamwise direction, i.e., $\lambda \mathbf{u}\mathbf{u}^T \mathbf{v} = 0$ for any $\mathbf{v}$ perpendicular to the flow $\mathbf{u}$.

The streamwise diffusion approach, in its many guises, has been exploited by many authors; see, e.g., Douglas and Russell [1982]; Hughes and Brooks [1979]; Johnson and Saranen [1986]; Johnson, Schatz, and Wahlbin [1987]; and Pironneau [1982].

For another, fully analyzed, finite element upwinding approach, see Girault and Raviart [1982 and 1986]. It should be pointed out that although seemingly quite different, many upwinding methods are actually very closely related.

Bibliography

Adams, R. [1975]. *Sobolev Spaces*, Academic, New York.

Arnold, D., Brezzi, F., and Fortin, M. [1984]. A stable finite element for the Stokes equations. *Calcolo* **21,** 337–344.

Axelsson, O. and Barker, V. [1984]. *Finite Element Solution of Boundary Value Problems*, Academic, Orlando.

Babuska, I. [1971]. Error bounds for finite element method. *Numer. Math.* **16,** 322–333.

Babuska, I. [1973]. The finite element method with Lagrange multipliers. *Numer. Math.* **20,** 179–192.

Babuska, I. [1988]. The p and h-p versions of the finite element method: The state of the art. *Finite Elements, Theory and Application* (Ed. by D. Dwoyer, M. Hussaini and R. Voigt), Springer, New York, 199–239.

Babuska, I. and Aziz, A. [1972]. Survey lectures on the mathematical foundations of the finite element method. *The Mathematical Foundations of the Finite Element Method with Application to Partial Differential Equations* (Ed. by A. Aziz), Academic, New York, 1–359.

Baiocchi, C. and Capelo, A. [1984]. *Variational and Quasivariational Inequalities*, Wiley, Chichester.

Baker, A. [1983]. *Finite Element Computational Fluid Mechanics*, Hemisphere, Washington.

Baker, G., Dougalis, V., and Karakashian, O. [1982]. On a higher order accurate fully discrete Galerkin approximation to the Navier–Stokes equations. *Math. Comp.* **39**, 339–375.

Bayliss, A., Gunzburger, M., and Turkel, E. [1982]. Boundary conditions for the numerical solution of elliptic equations in exterior regions. *SIAM J. Appl. Math.* **42**, 430–451.

Bégis, D. [1972]. Analyse numérique de l'écoulment d'un fluide de Bingham. *These*, Université Pierre et Marie Curie, Paris.

Bégis, D. and Glowinski, R. [1982]. Application de méthodes Lagrangien augmenté à la simulation numérique d'écoulment bidimensionnels de fluides visco-plastiques incompressibles. *Méthodes de Lagrangien Augmenté. Application à la Resolution Numérique de Problèmes aux Limites* (Ed. by M. Fortin and R. Glowinski), Dunod-Bordas, Paris, 219–240.

Bendali, A., Dominguez, J., and Gallic, S. [1985]. A variational approach for the vector potential formulation of the Stokes and Navier–Stokes equations in three dimensional domains. *J. Math. Anal. Appl.* **107**, 537–560.

Bercovier, M. [1978]. Perturbation of mixed variational problems. Application to mixed finite element methods. *RAIRO Anal. Numer.* **12**, 211–236.

Bercovier, M. and Pironneau, O. [1979]. Error estimates for finite element solution of the Stokes problem in primitive variables. *Numer. Math.* **33**, 211–224.

Berry, M., Heath, M., Kaneko, I., Lawo, M., Plemmons, R., and Ward, R. [1985]. An algorithm to compute a sparse basis of the null space. *Numer. Math.* **47**, 483–504.

Boland, J. and Nicolaides, R. [1983]. Stability of finite elements under divergence constraints, *SIAM J. Numer. Anal.* **20**, 722–731.

Boland, J. and Nicolaides, R. [1984]. On the stability of bilinear-constant velocity pressure finite elements. *Numer. Math.* **44**, 219–222.

Boland, J. and Nicolaides, R. [1985]. Stable and semistable low order finite elements for viscous flows. *SIAM J. Numer. Anal.* **22**, 474–492.

Bradshaw, P., Cebeci, T., and Whitelaw, J. [1981]. *Engineering Calculation Methods for Turbulent Flow*, Academic, New York.

Brezzi, F. [1974]. On the existence, uniqueness, and approximation of saddle-point problems arising from Lagrange multipliers. *RAIRO Anal. Numer.* **8**, 129–151.

Brezzi, F. and Douglas, J. [1988]. Stabilized mixed methods for the Stokes problem. *Numer. Math.* **53**, 225–235.

Brezzi, F. and Pitkaranta, J. [1984]. On the stabilization of finite element approximations of the Stokes problem. *Efficient Solutions of Elliptic Systems* (Ed. by W. Hackbusch), Vieweg, Braunschweig, 11–19.

Brezzi, F., Rappaz, J., and Raviart, P.-A. [1980]. Finite-dimensional approximation of nonlinear problems, Part I: Branches of non-singular solutions. *Numer. Math.* **36**, 1–25.

Bristeau, M., Glowinski, R., Mantel, B., Periaux, J., Perrier, P., and Pironneau, O. [1980a]. A finite element approximation of Navier–Stokes equations for incompressible viscous fluids. Iterative methods of solution. *Approximation Methods for the Navier–Stokes Problems* (Ed. by R. Rautmann), Springer, Berlin, 78–128.

Bristeau, M., Glowinski, R., Periaux, J., Perrier, P., and Pironneau, O. [1979]. On the numerical solution of nonlinear problems in fluid mechanics by least squares and finite element methods. (1) Least squares formulations and conjugate gradient solution of the continuous problem. *Comput. Meth. Appl. Mech. Engrg.* **17/18**, 619–657.

Bristeau, M., Glowinski, R., Periaux, J., Perrier, P., Pironneau, O., and Poirer, G. [1980b]. Application of optimal control and finite element methods to the calculation of transonic flows and incompressible flows. *Numerical Methods in Applied Fluid Mechanics* (Ed. by B. Hunt), Academic, London, 203–312.

Brooks, A. and Hughes, T. [1982]. Streamline upwind/Petrov Galerkin formulations for convection dominated flows. *Comput. Meths. Appl. Mech. Engrg.* **30**, 199–259.

Cantor, M. [1983]. Numerical treatment of potential type equations on R^n: Theoretical considerations. *SIAM J. Numer. Anal.* **20**, 72–85.

Cayco, M. [1985]. Finite element methods for the streamfunction formulation of the stationary Navier–Stokes equations. *PhD Thesis*, Carnegie Mellon University, Pittsburgh.

Cayco, M. and Nicolaides, R. [1986]. Finite element technique for optimal pressure recovery from stream function formulation of viscous flows, *Math. Comp.* **46**, 371–377.

Cayco, M. and Nicolaides, R. [1989]. Analysis of nonconforming stream function and pressure finite element spaces for the Navier–Stokes equations. To appear, *Comput. Math. Appl.*

Chorin, A. [1967]. A numericaol method for solving incompressible viscous flow problems. *J. Comput. Phys.* **2**, 12–26.

Christie, I., Griffiths, D., Mitchell, A., and Zienckiewicz, O. [1976]. Finite element methods for second-order differential equations with significant first derivatives. *Int. J. Numer. Meth. Engrg.* **10**, 1389–1396.

Chung, T. [1978]. *Finite Element Analysis in Fluid Mechanics*, McGraw-Hill, New York.

Ciarlet, P. [1978]. *The Finite Element Method for Elliptic Problems*, North-Holland, Amsterdam.

Conca, C. [1984]. Approximation de problèmes de type Stokes par élémentes finis mixtes. *Numer. Math.* **45**, 75–91.

Crochet, M., Davies, A., and Walters, K. [1984]. *Numerical Simulation of Non-Newtonian Flow*, Elsevier, Amsterdam.

Crochet, M. and Walters, K. [1983]. Numerical methods in non-Newtonian fluid mechanics. *Ann. Rev. Fluid Mech.* **15**, 241–260.

Crouzeix, M. and Raviart, P.-A. [1973]. Conforming and nonconforming finite element methods for solving the stationary Stokes equations. *RAIRO Anal. Numer.* **7**, 33–76.

Cuvelier, C. [1976]. Optical control of a system governed by the Navier–Stokes equations coupled with the heat equations. *New Developments in Differential Equations* (Ed. by W. Eckhaus), North-Holland, Amsterdam, 81–98.

Cuvelier, C. [1978]. Resolution numérique d'un problème de controle optimal d'un couplage des équations de Navier–Stokes et celle de la chaleur. *Calcolo* **15**, 345–379.

Cuvelier, C., Segal, A., and van Steenhoven, A. [1986]. *Finite Element Methods and Navier–Stokes Equations*, Reidel, Dordrecht.

den Heijer, C. and Rheinboldt, W. [1981]. On steplength algorithms for a class of continuation methods. *SIAM J. Numer. Anal.* **18**, 925–948.

Dennis, J. and More, J. [1977]. Quasi-Newton methods, motivations and theory. *SIAM Review* **19**, 46–89.

Dennis, J. and Schnabel, R. [1983]. *Numerical Methods for Unconstrained Optimization and Nonlinear Equations*, Prentice-Hall, Englewood Cliffs.

Douglas, J. and Russell, T. [1982]. Numerical methods for convection dominated diffusion problems based on combining the methods of characteristics and the finite element methods. *SIAM J. Numer. Anal.* **19**, 871–885.

Drazin, P. and Reid, W. [1981]. *Hydrodynamic Stability*, Cambridge, Cambridge.

Driessen, J. [1984]. Effects of non-uniform surface tension in fluid flow: Marangoni effect. *MS Thesis*, Delft University of Technology, Delft.

Du, Q. and Gunzburger, M. [1989a]. Analysis of Ladyzhenskaya model for incompressible viscous flow. To appear.

Du, Q. and Gunzburger, M. [1989b]. Finite element approximations of a Ladyzhenskaya model for stationary incompressible viscous flow. To appear.

Dukowicz, J. and Ramshaw, J. [1979]. Tensor viscosity method for convection in numerical fluid mechanics. *J. Comput. Phys.* **32**, 71–79.

Duvaut, G. and Lions, J.-L. [1976]. *Inequalities in Mechanics and Physics*, Springer, Berlin.

Engelman, M., Sani, R., and Gresho, P. [1982]. The implementation of normal and/or tangential boundary conditions in finite element codes for incompressible fluid flow. *Int. J. Numer. Meth. Fluids* **2**, 225–238.

Engelman, M., Strang, G., and Bathe, K.-J. [1981]. The application of quasi-Newton methods in fluid mechanics. *Int. J. Numer. Meth. Engrg.* **17**, 707–718.

Falk, R. [1975]. An analysis of the penalty method and extrapolation for the stationary Stokes problem. *Advances in Computer Methods for Partial Differential Equations* (Ed. by R. Vichnevetsky), AICA, New Brunswick, 66–69.

Fix, G.. and Gunzburger, M. [1977]. Downstream boundary conditions for viscous flow problems. *Comput. Math. Appl.* **3**, 53–63.

Fix, G., Gunzburger, M., and Nicolaides, R. [1981]. On mixed finite element methods for first-order elliptic systems. *Numer. Math.* **37**, 29–48.

Fix, G., Gunzburger, M., Nicolaides, R., and Peterson, J. [1984]. Mixed finite element approximations for the biharmonic equations, *Proc. 5th International Symposium on Finite Elements and Flow Problems* (Ed. by G. Carey and J. Oden), University of Texas, Austin, 281–286.

Fix, G., Gunzburger, M., and Peterson, J. [1983]. On finite element approximations of problems having inhomogeneous essential boundary conditions. *Comput. Math. Appl.* **9**, 687–700.

Fortin, M. [1972]. Calcul numérique des écoulmentes des fluides de Bingham et des fluides visqueux incompressibles par des méthodes d'éléments finis. *Thèse*, Université Pierre et Marie Curie, Paris.

Fortin, M. [1976]. Minimization of some non-differentiable functionals by the augmented Lagrangian method of Hestenes and Powell. *Appl. Math. Optim.* **2**, 236–250.

Fortin, M. [1977]. Analysis of the convergence of mixed finite element methods. *RAIRO Anal. Numer.* **11**, 341–354.

Fortin, M. and Fortin, A. [1985]. A generalization of Uzawa's algorithm for the solution of the Navier–Stokes equations. *Comm. Appl. Numer. Meth.* **1**, 205–208.

Fursikov, A. [1982]. Control problems and theorems concerning the unique solvability of a mixed boundary value problem for the three-dimensional Navier–Stokes and Euler equations. *Math. USSR Sbornik* **43**, 251–273.

Fursikov, A. [1983a]. On some control problems and results concerning the unique solvability of a mixed boundary value problem for the three-dimensional Navier–Stokes and Euler systems. *Soviet Math. Dokl.* **21**, 889–893.

Fursikov, A. [1983b]. Properties of solutions of some extremal problems connected with the Navier–Stokes system. *Math. USSR Sbornik* **46**, 323–351.

George, A. and Liu, J. [1981]. *Computer Solution of Large Sparse Positive Definite Systems*, Prentice-Hall, Englewood Cliffs.

Georgescu, A. [1985]. *Hydrodynamic Stability Theory*, Nijhoff, Dordrecht.

Ghia, U., Ghia, K., and Shin, C. [1982]. High-Re solution for incompressible viscous flow using the Navier–Stokes equations and a multigrid method. *J. Comput. Phys.* **48**, 387–395.

Girault, V. [1987]. Incompressible finite element methods for Navier–Stokes equations with non-standard boundary conditions in R^3. *Publication du Laboratoire d'Analyse Numerique* R87036, Universite Pierre et Marie Curie, Paris.

Girault, V. [1988]. Curl-conforming finite element methods for Navier–Stokes equations with non-standard boundary conditions in R^3. *Publication du Laboratoire d'Analyse Numerique* R88010, Universite Pierre et Marie Curie, Paris.

Girault, V. and Raviart, P.-A. [1979]. *Finite Element Approximation of the Navier–Stokes Equation*, Springer, Berlin.

Girault, V. and Raviart, P.-A. [1982]. An analysis of upwind schemes for the Navier–Stokes equations. *SIAM J. Numer. Anal.* **19**, 312–333.

Girault, V. and Raviart, P.-A. [1986]. *Finite Element Methods for Navier–Stokes Equations*, Springer, Berlin.

Glowinski, R. [1984]. *Numerical Methods for Nonlinear Variational Problems*, Springer, New York.

Glowinski, R., Lions, J.-L., and Trémolières, R. [1981]. *Numerical Analysis of Variational Inequalities*, North-Holland. Amsterdam.

Glowinski, R., Periaux, J., and Pironneau, O. [1980]. An efficient preconditioning scheme for iterative numerical solution of partial differential equations. *Appl. Math. Model.* **4**, 187–192.

Glowinski, R. and Pironneau, O. [1979]. On a mixed finite element approximation of the Stokes problem I. Convergence of the approximate solution. *Numer. Math.* **33**, 397–424.

Goldstein, C. [1981]. The finite element method with non-uniform mesh sizes for unbounded domains. *Math. Comp.* **36**, 387–404.

Golub, G. and van Loan, C. [1983]. *Matrix Computations*, Johns Hopkins, Baltimore.

Gresho, P. [1988]. The finite element method in viscous incompressible flow. *Report UCR*-99221, Lawrence Livermore National Laboratory, Livermore.

Gresho, P., Lee, R., and Sani, R. [1980]. On the time-dependent solution of the incompressible Navier–Stokes equations in two- and three-dimensions. *Recent Advances in Numerical Methods in Fluids* **1** (Ed. by C. Taylor and K. Morgan), Pineridge, Swansea, 27–80.

Gresho, P. and Sani, R. [1988]. On pressure boundary conditions for the incompressible Navier–Stokes equations, *Int. J. Numer. Meth. Fluids* **7**, 1111–1145.

Grisvard, P. [1985]. *Elliptic Problems in Nonsmooth Domains*, Pitman, Boston.

Guevremont, G., Habashi, W., Hafez, M., and Peeters, M. [1988]. A velocity–vorticity finite element formulation of the compressible Navier–Stokes equations. *Computer Mechanics '88: Theory and Applications. Proceedings of the International Conference on Computational Engineering Science* (Ed. by S. Atluri and G. Yagawa), Springer, Berlin, 51.x.1–51.x.4.

Guirguis, G. [1986]. On the existence, uniqueness, and regularity of the exterior Stokes problem in R^3. *Comm. Partial Diff. Eqns.* **11**, 567–594.

Guirguis, G. [1987a]. A third-order boundary condition for the exterior Stokes problem in three dimensions. *Math. Comp.* **49**, 379–389.

Guirguis, G. [1987b]. On the coupling of boundary integral and finite element methods for the exterior Stokes problem in 3-D. *SIAM J. Numer. Anal.* **24**, 310–322.

Guirguis, G. [1988]. On the competition between artificial methods and boundary integral methods for the Stokes equations in exterior domains. *Comm. Appl. Numer. Meth.* **4**, 491–497.

Guirgius, G. and Gunzburger, M. [1987]. On the approximation of the exterior Stokes problem in three dimensions. *Mod. Math. Anal. Numer.* **21**, 445–464.

Gunzburger, M., Hou, L., and Svobodny, T. [1989]. Analysis and finite element approximation of optimal control problems for the stationary Navier–Stokes equations. Part I: Neumann boundary and distributed controls; Part II: Dirichlet boundary controls; Part III: Constrained discrete Dirichlet boundary controls. To appear.

Gunzburger, M., Liu, C., and Nicolaides, R. [1983]. A finite element method for diffusion dominated unsteady viscous flows. *Comp. Meth. Appl. Mech. Engrg.* **39**, 55–67.

Gunzburger, M., Meir, A., and Peterson, J. [1989]. On the existence, uniqueness, and finite element approximation of the equations of stationary, incompressible MHD. To appear.

Gunzburger, M., Mundt, M., and Peterson, J. [1989]. Experiences with finite element methods for the velocity–vorticity formulation of three-dimensional viscous incompressible viscous flows. *Computational Methods for Viscous Flows* **4**, to appear.

Gunzburger, M. Nicolaides, R. [1984]. Issues in the implementation of substructuring algorithms for the Navier–Stokes equations. *Advances in Computer Methods for Partial Differential Equations* **V** (Ed. by R. Vichnevetsky and R. Stepleman), IMACS, New Brunswick, 57–63.

Gunzburger, M. and Nicolaides, R. [1985]. Elimination with noninvertible pivots. *Linear Algebra Appl.* **64**, 183–189.

Gunzburger, M. and Nicolaides, R. [1986]. On substructuring algorithms and solution techniques for the numerical approximation of partial differential equations. *Appl. Numer. Methods* **2**, 243–256.

Gunzburger, M., Nicolaides, R., and Liu, C. [1985]. Algorithmic and theoretical results on computation of incompressible viscous flows by finite element methods. *Comput. & Fluids* **13**, 361–373.

Gunzburger, M., Nicolaides, R., and Peterson, J. [1982]. On conforming finite element methods for incompressible viscous flows. *Comput. Math. Appl.* **8**, 167–179.

Gunzburger, M. and Peterson, J. [1983]. On conforming finite element methods for the inhomogeneous stationary Navier–Stokes equations, *Numer. Math.* **42**, 173–194.

Gunzburger, M. and Peterson, J. [1988]. Finite element methods for the streamfunction–vorticity equations: Boundary condition treatments and multiply connected domains. *SIAM J. Scient. Stat. Comput.* **9**, 650–668.

Gunzburger, M. and Peterson, J. [1989]. Predictor and steplength selection in continuation methods for the Navier–Stokes equations. To appear.

Gunzburger, M. and Turner, J. [1988]. An analysis of approximations of an algebraic model of turbulence. *Comput. Math. Appl.* **15**, 945–951.

Gustafson, K. and Hartman, R. [1983]. Divergence-free bases for finite element schemes in hydrodynamics, *SIAM J. Numer Anal.* **20**, 697–721.

Habashi, W., Peeters, M. Guevremont, G., and Hafez, M. [1987]. Finite element solutions of the compressible Navier–Stokes equations. *AIAA J.* **25**, 944–948.

Hafez, M., Habashi, M., Przybytkowski, S., and Peeters, M. [1987]. Compressible viscous internal flow calculations by a finite element method. *AIAA Paper 87-0644*, AIAA, New York.

Halim, A. and Hafez, M. [1984]. Calculation of separation bubbles using boundary layer type equations. *Computational Methods in Viscous Flows* **3**, Pineridge, Swansea, 395–415.

Hall, C., Peterson, J., Porsching, T., and Sledge, F. [1985]. The dual variable method for finite element discretizations of Navier/Stokes equations. *Int. J. Numer. Meth. Engrg.* **21**, 883–898.

Hanouzet, B. [1971]. Espaces de Sobolev avec poids. Application à un problème de Dirichlet dans un demi-éspace. *Rend. Sem. Mat. Univ. Padova* **46**, 227–272.

Heinrich, J., Huyakorn, P., Zienckiewicz, O., and Mitchell, A. [1977]. An upwind finite element scheme for the two-dimensional convective equation. *Int. J. Numer. Meth. Engrg.* **11**, 131–143.

Hinze, J. [1959]. *Turbulence*, McGraw Hill, New York.

Hughes, T. and Brooks, A. [1979]. A multidimensional upwind scheme with no crosswind diffusion. *Finite Element Methods for Convection Dominated Flow* (Ed. by T. Hughes), ASME, New York, 19–35.

Hughes, T. and Franca, L. [1987]. A new finite element formulation for computational fluid dynamics: VII. The Stokes problem with various boundary conditions: Symmetric formulations that converge for all velocity/pressure spaces. *Comput. Meths. Appl. Mech. Engrg.* **65**, 85–96.

Hughes, T., Franca, L., and Ballestra, M. [1986]. A new finite element formulation for computational fluid dynamics: V. Circumventing the Babuska–Brezzi condition: A stable Petrov–Galerkin formulation of the Stokes problem accommodating equal-order interpolations. *Comput. Meths. Appl. Mech. Engrg.* **59**, 85–99.

Hughes, T., Liu, W., and Brooks, A. [1979]. Finite element analysis of incompressible viscous flows by the penalty function formulation. *J. Comput. Phys.* **30**, 1–60.

Jamet, P. and Raviart, P.-A. [1973]. Numerical solution of the stationary Navier–Stokes equations by finite element methods. *Computing Methods in Applied Sciences and Engineering* (Ed. by R. Glowinski and J.-L. Lions), Springer, Berlin, 193–223.

Jean, M. and Pritchard, W. [1980]. The flow of fluids from nozzles at small Reynolds numbers. *Proc. Roy. Soc. London* **A370**, 61–72.

Jensen, S. and Vogelius, M. [1989]. Divergence stability in connection with the p-version of the finite element method. To appear.

Johnson, C. [1987]. *Numerical Solution of Partial Differential Equations by the Finite Element Method*, Cambridge, Cambridge.

Johnson, C. and Pitkaranta, J. [1982]. Analysis of some mixed finite element methods related to reduced integration. *Math. Comp.* **38**, 375–400.

Johnson, C. and Saranen, J. [1986]. Streamline diffusion methods for the incompressible Euler and Navier–Stokes equations. *Math. Comp.* **47**, 1–18.

Johnson, C., Schatz, A., and Wahlbin, L. [1987]. Crosswind smear and pointwise errors in streamline diffusion finite element methods. *Math. Comp.* **49**, 25–38.

Joseph, D. [1976]. *Stability of Fluid Motions*, Springer, Berlin.

Karakashian, O. [1982]. On a Galerkin–Lagrange multiplier method for the stationary Navier–Stokes equations. *SIAM J. Numer. Anal.* **19**, 909–923.

Karniadakis, G., Bullister, E., and Patera, A. [1986]. A spectral element method for solution of the two- and three-dimensional time-dependent incompressible Navier–Stokes equations. *Finite Element Methods for Nonlinear Problems: Proceedings of the Europe-US Symposium* (Ed. by P. Bergan, K.-J. Bathe and W. Wunderlich), Springer, Berlin, 803–817.

Keller, H. [1978]. Global homotopies and Newton methods. *Recent Advances in Numerical Analysis* (Ed. by C. de Boor and G. Golub), Academic, New York, 73–94.

Keller, H. [1987]. *Numerical Methods in Bifurcation Problems*, Springer, Berlin.

Kim, J. [1987]. On the initial-boundary value problem for a Bingham fluid in a three-dimensional domain. *Trans. AMS* **304**, 751–770.

Kim, J. [1989a]. Semi-discretization method for the three-dimensional motion of a Bingham fluid. To appear.

Kim, J. [1989b]. A finite element approximation of three-dimensional motion of a Bingham fluid. To appear.

Korczak, K. and Patera, A. [1986]. Isoparametic spectral element method for the solution of the Navier–Stokes equations in complex geometry. *J. Comput. Phys.* **62**, 361–382.

Ladyzhenskaya, O. [1969]. *The Mathematical Theory of Viscous Incompressible Flow*, Gordon and Breach, New York.

Ladyzhenskaya, O. [1970a]. New equations for the description of the viscous incompressible fluids and solvability in the large of the boundary value problems for them. *Boundary Value Problems of Mathematical Physics* **V** (Ed. by O. Ladyzhenskaya), AMS, Providence, 95–118.

Ladyzhenskaya, O. [1970b]. Modification of the Navier–Stokes equations for large velocity gradients. *Boundary Value Problems of Mathematical Physics and Related Aspects of Function Theory* **II** (Ed. by O. Ladyzhenskaya), Consultants Bureau, New York, 57–69.

Landau, L. and Lifshitz, E. [1987]. *Fluid Mechanics*, Pergamon, Oxford.

Lascaux, P. and Lesaint, P. [1975]. Some nonconforming finite elements for the plate bending problem. *RAIRO Anal. Numer.* **9**, 9–53.

Lee, R., Gresho, P., Chan, S., Sani, R., and M. Cullen [1982]. Conservation laws for primitive variable formulations of the incompressible flow equations using the Galerkin finite element method. *Finite Elements in Fluids* **4** (Ed. by R. Gallagher, D. Norrie, H. Oden and O. Zienkiewicz), Wiley, Chichester, 21–46.

LeRoux, M. [1977]. Méthode d'éléments finis pour la résolution numérique de problèmes extérieurs en dimension 2. *RAIRO Anal. Numer.* **11**, 27–60.

Lions, J.-L. [1969]. *Quelque Méthodes de Résolution des Problèmes aux Limites non Linéaires*, Dunod, Paris.

Lions, J.-L. [1985]. *Control of Distributed Singular Systems*, Bordas, Paris.

Mansfield, L. [1982]. Finite element subspaces with optimal rates of convergence for the stationary Stokes problem. *RAIRO Anal. Numer.* **16**, 49–66.

Marchal, J. and Crochet, M. [1987]. A new mixed finite element for calculating viscoelastic flow. *J. Non-Newtonian Fluid Mech.* **26**, 77–114.

Matthies, H. and Strang, G. [1979]. The solution of nonlinear finite element equations. *Int. J. Numer. Meth. Engrg.* **14**, 1613–1626.

Mercier, B. [1979a]. *Topics in Finite Element Solutions of Elliptic Problems*, Springer, Berlin.

Mercier, B. [1079b]. A conforming finite element method for two-dimensional incompressible elasticity. *Int. J. Numer. Meth. Engrg.* **14**, 942–945.

Mercier, B., Osborn, J., Rappaz, J., and Raviart, P.-A. [1981]. Eigenvalue approximation by mixed and hybrid methods. *Math. Comp.* **36**, 427–453.

Morgan, J. and Scott, R. [1975]. A nodal basis for C^1 piecewise polynomials of degree $n \geq 5$. *Math. Comp.* **29**, 736–740.

Nagtegaal, J., Parks, D., and Rice, J. [1974]. On numerically accurate finite element solutions in the fully plastic range, *Comp. Meth. Appl. Mech. Engrg.* **4**, 153–177.

Nécas, J. [1967]. *Les Méthodes Directes en Théorie des Equations Elliptiques*, Masson, Paris.

Nedelec, J. [1986]. A new family of mixed finite elements in $\mathbf{R}^3$. *Numer. Math.* **50**, 57–81.

Nedelec, J. and Planchard, J. [1973]. Une méthode variationnelle d'éléments finis pour la résolution numérique d'un problème extérieur dans $\mathbf{R}^3$. *RAIRO Anal. Numer.* **7**, 105–129.

Nicolaides, R. and Wu, X. [1988]. Applicability of nested dissection to two- and three-dimensional Navier–Stokes equations. *Computer Mechanics '88: Theory and Applications. Proceedings of the International Conference on Computational Engineering Science* (Ed. by S. Atluri and G. Yagawa), Springer, Berlin, 51.iv.1–51,iv.4.

Noor, A. [1980]. Reduced basis technique for nonlinear analysis of structures. *AIAA J.* **18**, 455–462.

Oden, J. and Reddy, J. [1976]. *An Introduction to the Mathematical Theory of Finite Elements*, Wiley, New York.

O'Leary, D. and Widlund, O. [1979]. Capacitance matrix methods for the Helmhotz equation on general three-dimensional regions. *Math. Comp.* **33**, 849–879.

Ortega, J. and Rheinboldt, W. [1970]. *Iterative Solution of Nonlinear Equations in Several Variables*, Academic, New York.

Osborn, J. [1976]. Approximation of the eigenvalues of a non-selfadjoint operator arising in the study of the stability of stationary solutions of the Navier–Stokes equations. *SIAM J. Numer. Anal.* **13**, 185–197.

Osborn, J. [1979]. Eigenvalue approximation by mixed methods. *Advances in Computer Methods for Partial Differential Equations* **III** (Ed. by R. Vichnevetsky and R. Stepleman), IMACS, New Brunswick, 158–161.

Osswald, G., Ghia, K., and Ghia, U. [1988]. Direct solution methodologies for the unsteady dynamics of an incompressible fluid. *Computer Mechanics '88: Theory and Applications. Proceedings of the International Conference on Computational Engineering Science* (Ed. by S. Atluri and G. Yagawa), Springer, Berlin, 51.vii.1–51.vii.4.

Patera, A. [1984]. A spectral element method for fluid dynamics: Laminar flow in a channel expansion. *J. Comput. Phys.* **54**, 468–488.

Peeters, M., Habashi, W., and Dueck, E. [1987]. Finite element stream function–vorticity solutions of the incompressible Navier–Stokes equations. *Int. J. Numer. Meth. Fluids* **7**, 17–27.

Pélissier, M. [1975]. Résolution numérique de quelques problèmes raides en mécanique des milieux faiblement compressible. *Calcolo* **12**, 275–314.

Peterson, J. [1980]. On eigenvalue approximations by mixed finite element methods. *PhD Thesis*, University of Tennessee, Knoxville.

Peterson, J. [1983]. An application of mixed finite element methods to the stability of the incompressible Navier–Stokes equation. *SIAM J. Sci. Stat. Comput.* **4**, 626–634.

Peterson, J. [1988]. On the finite element approximation of incompressible flows of an electrically conducting fluid. *Numer. Meth. Partial Diff. Eqns.* **4**, 57–68.

Peterson, J. [1989]. The reduced basis method for incompressible viscous flow calculations. To appear, *SIAM J. Sci. Stat. Comput.*

Peyret, R. and Taylor, T. [1983]. *Computational Methods for Fluid Flow*, Springer, New York.

Pironneau, O. [1982]. On the transport diffusion algorithm and its application to the Navier–Stokes equations. *Numer. Math.* **38**, 309–332.

Pironneau, O. [1984]. *Optimal Shape Design for Elliptic Systems*, Springer, New York.

Pironneau, O. [1986]. Conditions aux limites sur la pression pour les équations de Stokes et de Navier–Stokes. *C. R. Acd. Sci. Paris, Series I* **303**, 403–406.

Pitkaranta, J. [1982]. On a mixed finite element method for the Stokes problem in R^3. *RAIRO Anal. Numer.* **16**, 275–291.

Rheinboldt, W. [1980]. Solution fields of nonlinear equations and continuation methods. *SIAM J. Numer. Anal.* **17**, 221–237.

Roache, P. [1972]. *Computational Fluid Dynamics*, Hermosa, Albuquerque.

Ruas, V. [1985]. Quasisolenoidal velocity–pressure finite element methods for the three-dimensional Stokes problem. *Numer. Math.* **46**, 237–253.

Ruas, V. [1986]. Some nonstandard finite element methods for the numerical solution of viscous flow problems. *Tenth International Conference on Numerical Methods in Fluid Dynamics* (Ed. by F. Zhuang and Y. Zhu), Springer, Berlin, 538–544.

Sani, R., Gresho, P., Lee, R., Griffiths, D., and Engelman, M. [1981]. The cause and cure (?) of the spurious pressures generated by certain FEM solutions of the incompressible Navier–Stokes equations. *Int. J. Numer. Meth. Fluids* **1**, 17–43 and 171–204.

Schlicting, H. [1979], *Boundary Layer Theory*, McGraw-Hill, New York.

Schultz, M. [1973]. *Spline Analysis*, Prentice-Hall, Englewood Cliffs.

Scott, R. and Vogelius, M. [1985a]. Norm estimates for a maximal right inverse of the divergence operator in spaces of piecewise polynomials. *Math. Model. Numer. Anal.* **19**, 111–143.

Scott, R. and Vogelius, M. [1985b]. Conforming finite element methods for incompressible and nearly incompressible continua. *Large Scale Computations in Fluid Mechanics* **2** (Ed. by B. Engquist, S. Osher and R. Somerville), AMS, Providence, 221–244.

Segal, A. [1979]. On the numerical solution of the Stokes equations using the finite element method. *Comput. Meth. Appl. Mech. Engrg.* **19**, 165–185.

Segal, A. [1982]. Aspects of numerical methods for elliptic singular problems. *SIAM J. Sci. Stat. Comput.* **3**, 327–349.

Sequeira, A. [1983]. The coupling of boundary integral and finite element methods for the bidimensional exterior steady Stokes problem. *Math. Mech. Appl. Sci.* **5**, 356–375.

Sequeira, A. [1986]. On the computer implementation of a coupled boundary integral and finite element method for the bidimensional exterior steady Stokes problem. *Math. Meth. Appl. Sci.* **8**, 117–133.

Sermange, M. and Temam, R. [1983]. Some mathematical questions related to MHD equations. *Comm. Pure Appl. Math.* **36**, 635–664.

Serrin, J. [1958]. On the stability of viscous fluid motion. *Arch. Rational Mech. Anal.* **3**, 1–13.

Serrin, J. [1959]. Mathematical principles of classical fluid mechanics. *Encyclopedia of Physics* **VIII/1** (Ed. by S. Flugge and C. Truesdell), Springer, Berlin, 125–263.

Shercliff, J. [1965]. *A Textbook of Magnetohydrodynamics*, Pergamon, Oxford.

Sherif, A. and Hafez, M. [1983]. Computation of three-dimensional transonic flows using two stream functions. *AIAA Paper* 83-1948, AIAA, New York.

Silvester, D. and Thatcher, R. [1986]. The effect of the stability of mixed finite element approximations on the accuracy and rate of convergence of solutions when solving incompressible flow problems. *Int. J. Numer. Meth. Fluids* **6**, 841–853.

Stanisic', M. [1985]. *The Mathematical Theory of Turbulence*, Springer, Berlin.

Stenberg, R. [1984]. Analysis of mixed finite element methods for the Stokes problem: A unified approach. *Math. Comp.* **42**, 9–23.

Stenberg, R. [1987]. On some three-dimensional finite elements for incompressible media. *Comput. Meth. Appl. Mech. Engrg.* **63**, 261–269.

Strang, G. and Fix, G. [1973]. *An Analysis of the Finite Element Method*, Prentice-Hall, Englewood Cliffs.

Taylor, C. and Hood, P. [1973]. A numerical solution of the Navier–Stokes equations using the finite element method. *Comput. & Fluids* **1**, 73–100.

Temam, R. [1968]. Une méthode d'approximation de la solution des équations de Navier–Stokes. *Bull. Soc. Math. France* **96**, 115– 152.

Temam, R. [1979]. *Navier–Stokes Equations*, North-Holland, Amsterdam.

Temam, R. [1983]. *Navier–Stokes Equations and Nonlinear Functional Analysis*, SIAM, Philadelphia.

Tennekes, H. and Lumley, J. [1987]. *A First Course in Turbulence*, MIT, Cambridge.

Tezduyar, T., Glowinski, R., and Liou, J. [1988]. Petrov–Galerkin methods on multiply connected domains for the vorticity–stream function formulation of the incompressible Navier–Stokes equations. *Int. J. Numer. Meth. Fluids* **8**, 1269–1290.

Thatcher, R. and Silvester, D. [1987]. A locally mass conserving quadratic velocity, linear pressure element. *Numerical Analysis Report No. 147*, University of Manchester, Manchester.

Thomasset, F. [1981]. *Implementation of Finite Element Methods for Navier–Stokes Equations*, Springer, New York.

Tidd, D., Thatcher, R., and Kaye, A. [1986]. The free surface of Newtonian and non-Newtonian fluids trapped by surface tension. *Numerical Analysis Report No. 127*, University of Manchester, Manchester.

Turner, J. [1989]. On a zero equation model of turbulence. *Numer. Meth. Partial Diff. Eqns.* **5**, 25–33.

van Schaftingen, J. and Crochet, M. [1984]. A comparison of mixed methods for solving the flow of a Maxwell fluid. *Int. J. Numer. Meth. Fluids* **4**, 1065–1081.

Verfurth, R. [1984a]. Error estimates for a mixed finite element approximation of the Stokes equations. *RAIRO Anal. Numer.* **18**, 175–182.

Verfurth, R. [1984b]. A multilevel algorithm for mixed problems. *SIAM J. Numer. Anal.* **21**, 264–271.

Verfurth, R. [1985]. Finite element approximation of steady Navier–Stokes equations with mixed boundary conditions. *Mod. Math. Anal. Numer.* **19**, 461–475.

Verfurth, R. [1987a]. Finite element approximation of incompressible Navier–Stokes equations with slip boundary condition. *Numer. Math.* **50**, 697–721.

Verfurth, R. [1987b]. Mixed finite element approximation of the vector potential. *Numer. Math.* **50**, 685–695.

Verfurth, R. [1988]. A multilevel algorithm for mixed problems. II. Treatment of the mini-element. *SIAM J. Numer. Anal.* **25**, 285–293.

Vogelius, M. [1983]. A right-inverse for the divergence operator in spaces of piecewise polynomials. *Numer. Math.* **41**, 19–37.

Wait, R. and Mitchell, A. [1985]. *Finite Element Analysis and Applications*, Wiley, Chichester.

Wloka, J. [1987]. *Partial Differential Equations*, Cambridge, Cambridge.

Yanenko, N. [1971]. *The Method of Fractional Steps*, Springer, Berlin.

Zienkiewicz, O. [1977]. *The Finite Element Method*, McGraw-Hill, London.

Index of Symbol Definitions

Only those symbols that are used in chapters other than the one in which they are defined are listed. The definition of other symbols can be found somewhere within the chapter in which they are used. The numbers refer to the page on which the symbol is defined.

Author Index

Subject Index

Computer Science and Scientific Computing

Werner Rheinboldt and Daniel Siewiorek, editors

Allen B. Tucker, Jr., *Text Processing: Algorithms, Languages, and Applications*

Martin Charles Golumbic, *Algorithmic Graph Theory and Perfect Graphs*

Gabor T. Herman, *Image Reconstruction from Projections: The Fundamentals of Computerized Tomography*

Webb Miller and Celia Wrathall, *Software for Roundoff Analysis of Matrix Algorithms*

Ulrich W. Kulisch and Willard L. Miranker, *Computer Arithmetic in Theory and Practice*

Louis A. Hageman and David M. Young, *Applied Iterative Methods*

I. Gohberg, P. Lancaster, and L. Rodman, *Matrix Polynomials*

Azriel Rosenfeld and Avinash C. Kak, *Digital Picture Processing, Second Edition, Volume 1, Volume 2*

Dimitri P. Bertsekas, *Constrained Optimization and Lagrange Multiplier Methods*

James S. Vandergraft, *Introduction to Numerical Computations, Second Edition*

Götz Alefeld and Jürgen Herzberger, *Introduction to Interval Computations*. Translated by Jon Rokne

Françoise Chatelin, *Spectral Approximation of Linear Operators*

Robert R. Korfhage, *Discrete Computational Structures, Second Edition*

Martin D. Davis and Elaine J. Weyuker, *Computability, Complexity, and Languages: Fundamentals of Theoretical Computer Science*

Leonard Uhr, *Algorithm–Structured Computer Arrays and Networks: Architectures and Processes for Images, Percepts, Models, Information*

O. Axelsson and V. A. Barker, *Finite Element Solution of Boundary Value Problems: Theory and Computation*

Philip J. Davis and Philip Rabinowitz, *Methods of Numerical Integration, Second Edition*

Norman Bleistein, *Mathematical Methods for Wave Phenomena*

Robert H. Bonczek, Clyde W. Holsapple, and Andrew B. Whinston, *Micro Database Management: Practical Techniques for Application Development*

Peter Lancaster and Miron Tismenetsky, *The Theory of Matrices: With Applications, Second Edition*

Peter B. Andrews, *An Introduction to Mathematical Logic and Type Theory: To Truth through Proof*